Environmental Science and Management

Environmental Science and Management

Abhijit Mallick, Ph.D.
Department of Chemistry
Academy of Technology, Hooghly,
West Bengal, India

Viva Books

New Delhi | Mumbai | Chennai | Kolkata | Bengaluru | Hyderabad | Kochi | Guwahati

Publisher's note

Every possible effort has been made to ensure that the information contained in this book is accurate at the time of going to press, and the publisher and author cannot accept responsibility for any errors or omissions, however caused. No responsibility for loss or damage occasioned to any person acting, or refraining from action, as a result of the material in this publication can be accepted by the editor, the publisher or the author.

Every effort has been made to trace the owners of copyright material used in this book. The author and the publisher will be grateful for any omission brought to their notice for acknowledgement in the future editions of the book.

First Published 2014

Viva Books Private Limited

- 4737/23, Ansari Road, Daryaganj, New Delhi 110 002
  Tel. 011-42242200, 23258325, 23283121, Email: vivadelhi@vivagroupindia.net
- 76, Service Industries, Shirvane, Sector 1, Nerul, Navi Mumbai 400 706
  Tel. 022-27721273, 27721274, Email: vivamumbai@vivagroupindia.net
- Megh Tower, Old No. 307, New No. 165, Poonamallee High Road, Maduravoyal, Chennai 600 095
  Tel. 044-23780991, 23780992, 23780994, Email: vivachennai@vivagroupindia.net
- B-103, Jindal Towers, 21/1A/3 Darga Road, Kolkata 700 017
  Tel. 033-22816713, Email: vivakolkata@vivagroupindia.net
- 7, Sovereign Park Aptts., 56-58, K. R. Road, Basavanagudi, Bengaluru 560 004
  Tel. 080-26607409, Email: vivabangalore@vivagroupindia.net
- 101-102, Moghal Marc Apartments, 3-4-637 to 641, Narayanguda, Hyderabad 500 029
  Tel. 040-27564481, Email: vivahyderabad@vivagroupindia.net
- First Floor, Beevi Towers, SRM Road, Kaloor, Kochi 682 018
  Tel. 0484-2403055, 2403056, Email: vivakochi@vivagroupindia.net
- 232, GNB Road, Beside UCO Bank, Silpukhuri, Guwahati 781 003
  Tel. 0361-2666386, Email: vivaguwahati@vivagroupindia.net

www.vivagroupindia.com

ISBN: 978-81-309-2490-8

Published by Vinod Vasishtha for Viva Books Private Limited, 4737/23, Ansari Road, Daryaganj, New Delhi 110 002.

Printed and bound by Raj Press, R-3, Inderpuri, New Delhi 110 012.

SUB-0001/SSUB-0005

Dedicated to

My beloved professor, Dr. Sadhan Kumar De, ex-professor, IIT (Kharagpur)

Contents

Acknowledgements

I am very much obliged to my family members including my father, my wife and my beloved son for their inspiration and continuous support without which it would have been impossible to complete this uphill task.

I am indebted to Viva Books Private Limited for their continuous support and inspiration to complete this book.

I must not miss this opportunity to acknowledge Prof. Mr. Jagannath Banerjee, Chairman, Academy of Technology and Prof. Ms. Anindita Banerjee, Executive Trustee of Academy of Technology, for their all-round cooperation and support.

I am also indebted to Dr. Dilip Bhattacharya, Director, Academy of Technology, for his cooperation and support.

I must not miss the opportunity to acknowledge my colleagues in the Department of Chemistry and Mr. Sadhan Das, Laboratory Assistant in the Department of Chemistry for their cooperation and support.

Abhijit Mallick

Preamble

This book is primarily written to focus on our environment. This book covers environmental chemistry, pollution and management to control abatement of pollution. So this book is expected to be suitable to all students including engineering, science and management.

Environment provides all basic needs to us for our survival. All of us should have sufficient knowledge and importance of our environment. So it is everybody's responsibility to take care of the environment.

The objective of this book is to discuss briefly about different types of environmental pollution, type of pollutants and their adverse effects on human being and overall ecosystem. Factory exhaust and automobile exhaust are the major sources of air pollution. Possible control measures to minimize the air pollution are well covered in this book.

Water pollution and filling up ponds and wetlands are major problems in our society. Details of water pollution, water pollutants and water treatment process are well documented in this book.

Rapid population growth creates different environmental problems. So study of human population dynamics is very important. The book also deals with different models to explain population growth. Mathematical details and numerical problems are also discussed.

Now-a-days two major problems are ozone layer depletion in stratosphere and rapid increase in greenhouse gases (GHG) in the Earth's atmosphere. Increase in GHG concentration in atmosphere results in global warming and unwanted atmospheric changes. The book covers details of these problems as well as possible control measures to minimize them.

Depletion of non-renewable resources is another major problem. Resource consumption and sustainable development are well documented in this book.

The book covers details of ISO concept of environmental management system or EMS, covering environmental audit, technology diffusion, LCA model and EIA.

Details of hardness of water, i.e., radicals responsible for hardness, problems arising due to hardness and also different softening processes, is also discussed in this book.

Prof. Abhijit Mallick
(dramallick@rediffmail.com)

Abbreviations

UNCED	United Nations Conference on Environment and Development
UNEP	United Nations Environment Programme, founded in 1972
GDP	Gross Domestic Product
CPCB	Central Pollution Control Board
EPA	Environmental Protection Agency
EIA	Environmental Impact Assessment
MSB	Marginal Social Benefit
MPC	Marginal Private Cost
NAAQS	National Ambient Air Quality Standards
NEPA	National Environmental Policy Act
CAA	Clean Air Act
IUCN	International Union for Conservation of Nature (1948)
OECD	Organisation for Economic Co-operation and Development
PSI	Pollution Standard Index
GATT	General Agreement on Trade and Tariff
IPR	Intellectual Property Right
IGBP	International Geosphere-Biosphere Programme
UNFCCC	United Nations Framework Convention on Climate Change
RCRA	Resource Conservation and Recovery Act
ZPG	Zero Population Growth
TFR	Total Fertility Rate

OTEC	Ocean Thermal Energy Conversion
CITES	Convention on International Trade in Endangered Species
UNESCO	United Nations Educational, Scientific and Cultural Organization
SPM	Suspended Particulate Matter
CNG	Compressed Natural Gas
IPCC	Intergovernmental Panel on Climate Change
OSHA	Occupational Safety and Health Administration
CHABA	Committee on Hearing, Bioacoustics and Biomechanics
CoR	Club of Rome
MSW	Municipal Solid Waste
GHG	Greenhouse Gas
DALR	Dry Adiabatic Lapse Rate
SALR	Saturated Adiabatic Lapse Rate
ELR	Environmental Lapse Rate
LFC	Level of Free Convection
LI	Lifted Index
LCL	Lifting Condensation Level
MSY	Maximum Sustainable Yield
ESP	Electrostatic Precipitator
LCA	Life-Cycle Assessment
TEL	Tetraethyl lead

1

Ecology

1.1 INTRODUCTION

1.1.1 Environmental science

(a) It is the science of chemical and biochemical phenomena constantly occurring in the nature.

(b) It is the study of sources, reactions, transport, effect and fate of a species in air, water and soil.

(c) It is the study of effect of human activities on different species and also on environment.

1.1.2 Goals or objectives

The main objective is to educate people regarding the following

(a) Importance and usefulness of environment.

(b) Protection and conservation of environment and natural resources.

(c) The need to restrict human activities which lead to indiscriminate release of pollutants in the environment.

1.2 ECOLOGY

It is the study of influence of environmental factors on different aspects of life, e.g., morphology, physiology, growth, distribution, behaviour and survival of living organisms. It is concerned with relationship between living organisms and their physical environment.

1.3 ECOSYSTEM

It is the study of interaction between biotic and abiotic components of the environment through exchange of matter and energy.

1.3.1 Characteristics of ecosystem

(a) An ecosystem contains two components: biotic and abiotic.
(b) All the nutrient cycles are completed through interactions between biotic and abiotic components.
(c) An ecosystem has no boundary. It is flexible.
(d) Adaptation to the new environment is a must for each and every species. Otherwise, the species will be destroyed.
(e) An ecosystem always passes from less complex state to more complex state. The phenomenon is called succession.
(f) *Natural ecosystem* The ecosystem which is created and maintained by nature itself is called natural ecosystem. Examples are forests, ponds, lakes etc.
(g) *Artificial ecosystem* The ecosystem which is built up and maintained by human beings is called artificial ecosystem. Examples are parks, zoological garden, aquarium etc.
(h) *Incomplete ecosystem* The ecosystem where all the components of environment are not present is called incomplete ecosystem. Examples are interior of caves, bottom of ocean etc., where producers are absent.
(i) *Biome* The major ecological community, occupying a very large area, is called biome. Examples are rainforest, ocean, desert etc.

1.3.2 Components of ecosystem

(a) Abiotic components which are also known as environmental factors or abiotic factors.
(b) Biotic components.

1.3.2.1 Abiotic factors

The different environmental conditions, which have significant influence on existence of biotic components, are called abiotic factors. Abiotic factors are divided into different categories

(i) *Physical factors* Examples are: sunlight, wind, water, flood, temperature, pressure, humidity, rainfall etc.
(ii) *Inorganic factors* Examples are minerals, oils, fuels, gases etc.
(iii) *Organic factors* Examples are carbohydrates, proteins, lipids etc.

1.3.2.2 Biotic components

Biotic components are mainly divided into two parts

(i) *Autotrophs* Green plants are producing their own food through photosynthesis. These are called autotrophs.
(ii) *Heterotrophs* Besides green plants other living organisms depend on green plants for their survival. These are called heterotrophs. These heterotrophs are again divided into two categories
(a) Phagotrophs and (b) Saprotrophs.

(a) *Phagotrophs* The living organisms, collecting their food directly from the green plants, are called phagotrophs. These living organisms are again divided into three categories

1. **Herbivores or primary consumers** There are some animals, which consume green plants only for their survival. These animals are called herbivores. Examples are cow, goat, deer etc.
2. **Carnivores or secondary consumers** There are some animals, which consume herbivores and other animals for their survival. These are called carnivores. Examples are tiger, lion etc.
3. **Omnivores or tertiary consumers** There are some living organisms, which consume both green plants and other animals for their survival. These are called omnivores. Examples are human being etc.

(b) *Saprotrophs* The organisms, which survive only on dead animals and plants, are called saprotrophs. Examples: fungi, protozoa, bacteria etc.

1.3.3 Aquatic ecosystem

An aquatic ecosystem is the way in which living organisms interact within a body of water. Essentially it describes the relationship between plants and animals within an aquatic environment. These systems exist in places such as oceans, ponds, lakes, streams and rivers. In any aquatic ecosystem there are three components

(a) *Bottom layer* Bottom of any aquatic ecosystem is made of soil, sand, rocks and minerals. High water pressure exits near the bottom due to which animals, residing in this region are flat.

(b) *Water body* Next layer is the water itself, which could be freshwater or saltwater and static or moving. For example a river flows but a swamp is stagnant and does not move. This layer is the habitat of different aquatic living organisms. A food chain, consisting of different types of consumers, exists in this layer. Aquatic plants are the main autotrophs of aquatic ecosystem. These plants are crucial to the ecosystem for re-oxygenating water and providing a habitat to creatures. Aquatic animals vary widely in size and type from crustaceans to whales. Primary consumers are insects, which survive on green plants. Secondary consumers are small fishes, which consume insects. Large fishes and seals, which consume small fishes, are the next higher predators. Sharks and whales are the larger predators.

(c) *Top layer* Top layer of an aquatic ecosystem is the water surface in touch with the atmosphere, which supplies necessary oxygen and other minerals to the aquatic ecosystem to maintain its clarity and quality.

1.3.4 Mangrove ecosystem

1.3.4.1 General description

There are some highly specialized plants such as palms, ferns, vines, shrubs etc., which are usually found in waterlogged saline soils subjected to regular flooding. The ecosystem consisting of these plants is called mangrove ecosystem. This type of ecosystem is commonly found within tropical and subtropical sheltered coastal areas subjected to tidal influences. Wide variety of species and broad species diversity is the main criteria of mangrove ecosystem.

The species in mangrove ecosystem can be divided into two broad categories: (a) exclusive and (b) non-exclusive. There are 60 different plant species in exclusive mangrove ecosystem, which is confined to intertidal areas and not found within any other type of vegetation community. There are 20 plant species in non-exclusive type mangrove ecosystem. This type of ecosystem is found not only in intertidal areas but also in other terrestrial areas.

1.3.4.2 Requirements for growth of mangrove ecosystem

Mangroves are commonly found throughout the world between latitudes 32°N and 38°S. Following are some requirements for growth of mangrove ecosystem

(a) *Temperature* Mangrove ecosystem is generally found in those coastal areas where temperature lies within 10ºC and 20ºC. Species diversification is limited if the temperature falls below or rises above.

(b) *Rainfall* Mangrove ecosystem grows in those areas which receive heavy rainfall with sufficient run off.

(c) *Protection* Mangrove ecosystem needs protection from strong winds and tidal waves. This is the reason for which mangrove ecosystem is located within sheltered coastal areas, surrounded by indented estuaries and offshore islands. One of such mangrove ecosystem is shown in the following figure.

Fig. 1.1 A scenario of mangrove ecosystem.

1.3.4.3 Beneficial effects of mangrove ecosystem

(a) Several types of forestry products (firewood, charcoal, timber, honey etc.) and fishery products (fish, prawn, crab, mollusk etc.) are obtained from mangrove ecosystem. The mangrove wood with high content of tannin is durable and hence used as timber.

(b) Mangrove extracts are very much useful and used in manufacturing indigenous medicine.
   (i) **Bruguiera** leaves are useful to reduce blood pressure.
   (ii) **Excoecaria agallocha** is used for the treatment of leprosy and epilepsy.
   (iii) Roots and stems of **Derris trifoliata** are used for narcotizing fishes.
   (iv) **Acanthus ilicifolius** is used in the treatment of rheumatic disorders.

(c) (i) Mangroves protect coastal areas from UV-B radiation.
   (ii) Mangroves remove huge amount of CO_2 from the atmosphere through photosynthesis, thereby reducing the global warming problem.
   (iii) Mangrove forests protect coastal communities from the fury of cyclones, storms, floods and tsunami. In India devastating tsunami rocked Andaman and Nicobar Islands and southeast coast of the country but the areas surrounded by mangrove ecosystem remained unaffected.
   (iv) Mangroves protect the underground water, which is a very important source of drinking water supply to coastal population, by preventing seawater from entering into the island.
   (v) The mangrove systems minimise the action of waves and thus prevent the coast from erosion.

1.4 SOME DEFINITIONS

1.4.1 GPP

In a given ecosystem the total amount of solar energy converted into chemical energy by green plants through photosynthesis is called gross primary production or GPP. Factors influencing GPP are (a) the light intensity, (b) the quantum efficiency of light, (c) the leaf area index (LAI), (d) the leaf arrangement on plants, (e) the type of plant.

LAI It is the area of leaf surface exposed to sunlight over a specified area of ground surface.

1.4.2 NPP

In a given ecosystem green plants are responsible for GPP, which is essential for survival of life on Earth. A fraction of GPP is utilised by green plants for their survival and the other fraction, which is available to transfer to other living organisms, is called net primary production or NPP.

1.4.3 Niche

Ecological niche is a very basic and fundamental ecological concept.

(a) It defines the position of a species within an ecosystem.
(b) It describes both the range of conditions necessary for survival of the species and ecological role, played by the species, in that ecosystem.
(c) Ecological niche subsumes types of interactions between a species and its surroundings including the biotic and abiotic components.

(d) If the concept of ecological niche is based on that set of conditions within which an organism lives without competition for food, it is called fundamental niche.

(e) If the concept of ecological niche is based on that set of conditions within which an organism lives with severe competition with others for food, it is called realized niche.

1.4.4 Habitat

A habitat (a Latin word, which means "it inhabits") is an ecological or environmental area that is inhabited by a particular species of animal, plant or other type of organism. It is the natural environment in which an organism lives, or the physical environment that surrounds (influences and is utilized by) a species population. The habitat must supply the needs of organisms, such as food, water, temperature, oxygen, and minerals.

Microhabitat The term microhabitat is often used to describe the small scale physical requirements of a particular organism or population.

Biologically, the habitat is the organism's "address" and the niche is its "profession".

1.4.5 Plankton

Hansen (1937) coined the term plankton to apply to minute floating organisms, usually found in surface water. The movement of plankton depends on the direction of water current. Plankton group can be classified into different categories as under

Category	*Size range*
Megaplankton	> 20 mm
Macroplankton	2 to 20 mm
Mesoplankton	0.2 to 2 mm
Microplankton	20 to 200 μm
Nanoplankton	2 to 20 μm
Picoplankton	0.2 to 2 μm
Femtoplankton	< 0.2 μm

1.4.6 Biomagnification

In any given ecosystem some chemicals are continuously accumulated in living organisms. ***Biomagnification (or bioaccumulation)*** refers to the ability of living organisms to accumulate certain chemicals to a concentration higher than that present in the environment. Generally organisms accumulate different chemical substances for their nutrition. In environmental science, however, the major focus of biomagnification is the accumulation of chlorinated hydrocarbons, such as DDT and PCBs, which are insoluble in water but highly soluble in fats. Because almost all fats within ecosystems occur in the living bodies of organisms, these chlorinated hydrocarbons tend to selectively accumulate in organisms. This can lead to ecological imbalance. Accumulation of a substance only through contact with water is known as ***bioconcentration***.

1.4.7 Biodiversity

Each and every ecosystem possesses several types of species, including producers, consumers of different trophic levels. Environmental condition has been changing continuously since the formation of the Earth, as a result of which, many species became extinct and some other species have appeared. This natural diversion of species is known as biodiversity.

1.4.8 Receptor

The organism, which receives pollutants and is affected by those pollutants, is called receptor. Example: human being.

1.4.9 Sink

The material, which can retain pollutants for long period and gets interacted with them, is called sink. One example is marble.

1.4.10 Speciation

A pollutant may exist in different forms. The method of determination of different forms of a pollutant is called speciation. For example, mercury exists as elemental Hg and also as CH_3Hg^+.

1.4.11 Bioremediation

It refers to the use of microbes such as bacteria and some fungi to break down pollutants into less toxic and simple organics. Microbes either consume or break down human wastes, pollutants, oil, pesticides and toxic wastes such as cyanides and chlorinated hydrocarbons.

1.4.12 Flora and Fauna

In a given ecosystem, ***flora*** refers to plants and ***fauna*** refers to animals.

1.5 FOOD CHAIN

Autotrophs are called producers, which convert solar energy into chemical energy. Consumers collect energy through consumption of autotrophs. So all the living organisms and green plants are interconnected through different chains, known as food chains. Food chains may be of different types

(a) *Predator food chain* The food chain, which starts with producer and ends to the highest trophic level, is called predator food chain. One example is
Grass $\longrightarrow$ Grasshopper $\longrightarrow$ toad $\longrightarrow$ snake $\longrightarrow$ peacock $\longrightarrow$

(b) *Parasitic food chain* The food chain, which ends with parasitic animals, is called parasitic food chain. One example is
Man $\longrightarrow$ Worm $\longrightarrow$ Protozoa

(c) *Saprophytic food chain* The food chain, which starts with dead plant/animal, is called saprophytic food chain. One example is
Dead plant/animal ⟶ Fungi ⟶ Bacteria

An example of **desert food chain** is shown below
Producers ⟶ Primary consumers ⟶ Secondary consumers ⟶ Larger predators
Producers Cactus, Annual flowers, Creosote bushes.
Primary consumers Insects, Small Lizards, Rodents, Kangaroo, Rats.
Secondary consumers Tarantulas, Scorpions, Large Lizards, Snakes.
Larger producers Hawk, Kit Fox.

A **grazing food chain** cycle of a given ecosystem is shown in the Fig. 1.2.

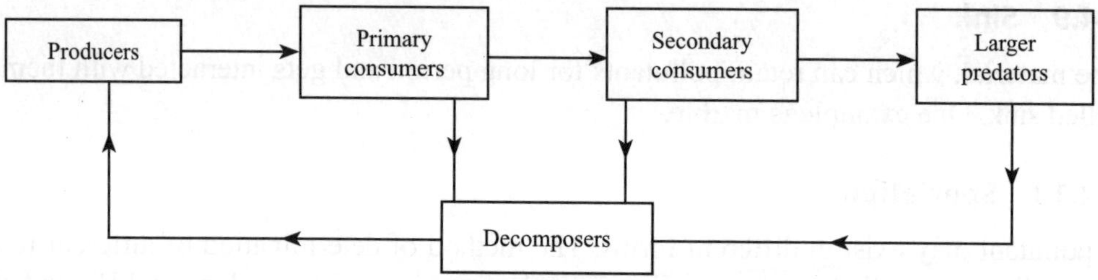

Fig. 1.2 Schematic diagram of a grazing food chain cycle.

1.5.1 Characteristics of a food chain

(a) In a given food chain, producers and consumers can be divided into different layers, known as trophic levels.
(b) As we move from lower trophic level to higher trophic level, size of organism increases but the number decreases. If all the trophic levels are sequentially placed one after another, a pyramid is formed. This pyramid is called **ecological pyramid**, shown in the Fig. 1.3.

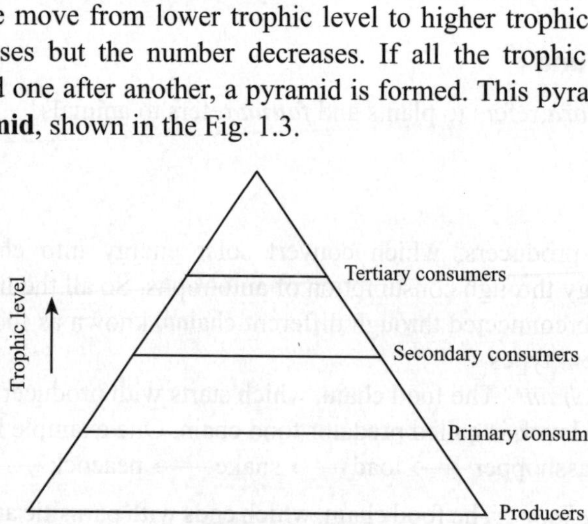

Fig. 1.3 Ecological pyramid.

(c) Energy is always flowing from producer to the higher trophic level and the flow of energy is **unidirectional**. However, flow of nutrients is cyclic in nature.

(d) In any given ecosystem there are several food chains and these food chains have common consumers. Hence these food chains are interconnected, forming a web, known as **food web**.

The following figure, Fig. 1.4, illustrates the energy transfer from producers to higher trophic levels.

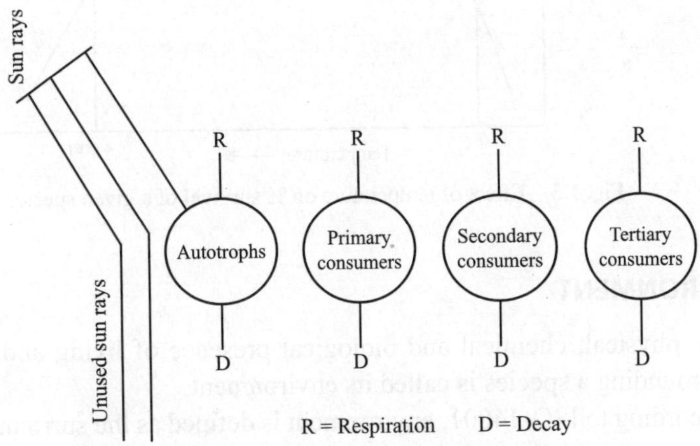

Fig. 1.4 Energy flow diagram of a given ecosystem.

1.6 ENVIRONMENTAL STRAIN AND STRESS

1.6.1 Plastic environmental strain

Adaptations, which give rise to plastic resistance to organism and thereby prevent injury by environmental stress are termed resistance adaptations, also known as plastic environmental strain.

1.6.2 Elastic environmental strain

There are some adaptations, which not only prevent injury by external stress but also permit growth of organism. These are called capacity adaptations, also known as elastic environmental strain.

1.6.3 Environmental stress

Any environmental factor, potentially unfavourable to living organisms, is called environmental stress. Survival of an organism depends on several abiotic factors. Each and every factor has two limits within which living organisms survive smoothly. One of such environmental factor is temperature. Living organisms lead a smooth life within the temperature limits 0ºC to +50ºC. The region is known as biokinetic zone. Region at either end of this zone is called zone of resistance. It is well illustrated in the Fig. 1.5.

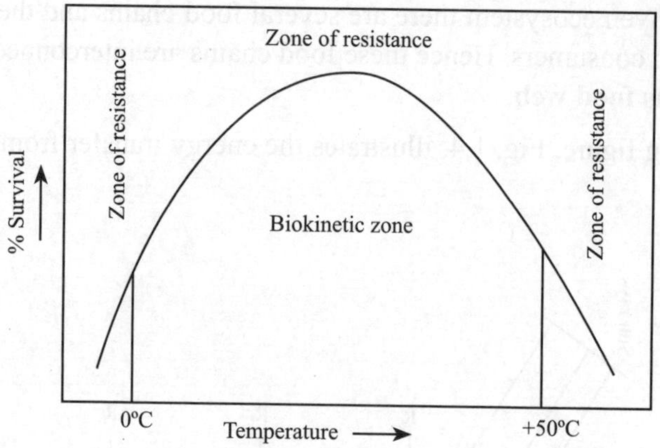

Fig. 1.5 Effect of temperature on % survival of a given species.

1.7 ENVIRONMENT

(a) The physical, chemical and biological presence of living and non-living things surrounding a species is called its environment.
(b) According to ISO-14001, environment is defined as the surroundings, comprising of air, water, land, natural resources and their interactions.

1.7.1 Classification

In one angle environment can be classified into four categories: (a) atmosphere, (b) hydrosphere, (c) lithosphere, (d) biosphere. An equilibrium is established between biosphere and the remaining three as follows:

$$\text{Biosphere} \rightleftharpoons \text{Atmosphere} + \text{Hydrosphere} + \text{Lithosphere}$$

Whenever the equilibrium is disturbed the environment is said to be polluted.

1.7.2 Causes of disturbance

(a) Civilisation growth, (b) Industrial growth, (c) Unplanned urbanisation, (d) Population explosion, (e) Lack of social responsibilities.

1.8 POLLUTION

(a) Sufficient alteration of natural environment (atmosphere, hydrosphere and lithosphere), which has deleterious effect on human beings and animals, is termed pollution. However, some degree of alteration is necessary for human activities but it has a definite tolerance limit. Whenever this limit is crossed the environment is said to be polluted.

(b) Multifarious activities of human being, like industrialization, construction, transportation, using chemicals in agriculture etc. lead to generation and release of objectionable matters to the environment, as a result of which, environment becomes hostile to the living organisms. It is called environmental pollution.

1.8.1 Threshold Limit Value (TLV)

For a given pollutant, this is a unique measure of toxicity, which indicates the conditions under which a worker is continuously exposed in a toxic environment without any adverse effect.

TLV is the maximum possible average concentration limit of a toxic pollutant under which a worker can work safely for 40 hours in a week. For a given pollutant, higher the TLV, less the toxicity. TLV values of some elements and gases are given in Table 1.1 and Table 1.2 respectively.

Table 1.1 TLV of different elements.

Element	*TLV ($\mu g/m^3$)*
Sn	2000
Cu	1000
As	500
Sb	500
Zn	500
Cd	200
Pb	200
Hg	10

Table 1.2 TLV of different gases.

Pollutant	*TLV(ppm)*
CO	50
NH_3	50
SO_2	5
NO_2	5
Cl_2	1
$COCl_2$	0.1
O_2	0.1
F_2	0.1
Br_2	0.1

Significance It is obvious from the above tables that higher the TLV of a pollutant, less harmful the pollutant is.

Unit of TLV The unit of TLV is usually expressed either in *ppm/ppb* or in mg/m^3 and $\mu g/m^3$. Both units are generally used in case of gaseous pollutants but *ppm/ppb* is frequently used in case of solid pollutants. In case of gaseous pollutants the relation between these two units is shown below

All gaseous pollutants are assumed as ideal gases. So, according to ideal gas relation, we have

$$PV = nRT = \frac{w}{M}RT$$

Or, $$\text{Density }(\rho) = \frac{w}{V} = \frac{PM}{RT}\ g/lit$$

Where, M represents molecular weight of the substance or pollutant. At 25°C, the value of RT/P is 22.45 lit.

So, $$\text{density }(\rho) = \frac{M}{22.45}\ g/lit = \frac{M \times 10^3}{22.45}\ g/m^3$$

So, $$y\ ppb = \frac{y \text{ gm of substance}}{10^9\ g \text{ of air}}$$

$$= \frac{y \text{ gm of substance}}{10^9\ g \text{ of air}} \times \frac{M \times 10^3}{22.45}\ g/m^3 = \frac{y \times M}{22.45}\ \mu g/m^3$$

Similarly, $$y\ ppm = \frac{y \times M}{22.45}\ mg/m^3$$

1.8.2 Pollutant

Any species/substance/gas/liquid, which remains in the environment with concentration more than its TLV, is known as pollutant.

1.9 POLLUTION STANDARD INDEX (PSI)

Environmental Protection Agency (EPA) has developed this standard to quantify the harmful effect of six criteria pollutants.

(i) According to this standard if PSI of a given pollutant reaches 100, the pollutant is assumed to reach its TLV.

(ii) When PSI of the said pollutant reaches 200, the environment is deteriorated and people with existing heart and lung patients are advised to stay indoors.

(iii) When PSI of the said pollutant reaches 300, the environment is deteriorated significantly. Power plants and other industries are advised to limit their operations.

(iv) When PSI of the said pollutant reaches 400, the environment is deteriorated to the alarming level. Most industrial activities are prohibited.

Table 1.3 PSI values, indicating different environmental conditions.

PSI	*Environment*
0 to 50	Good
51 to 100	Moderate
101 to 199	Unhealthful
200 to 299	Very unhealthful
≥ 300	Hazardous

1.10 BIOGEOCHEMICAL CYCLE

Principal nutrients, like carbon, nitrogen, sulphur and phosphorus, are constantly absorbed by living organisms from their environment and at the same time the nutrients are returned to the atmosphere through various biochemical reactions. This cyclic flow of nutrients from abiotic components to biotic components and back to abiotic components is called biogeochemical cycle. The principal nutrient cycles are

(a) Nitrogen cycle. It is shown in the Fig. 1.6.
(b) Carbon dioxide cycle. It is well illustrated in the Fig. 1.7.
(c) Sulphur cycle. It is shown in the Fig. 1.8.
(d) Phosphorus cycle. It is illustrated in the Fig. 1.9.

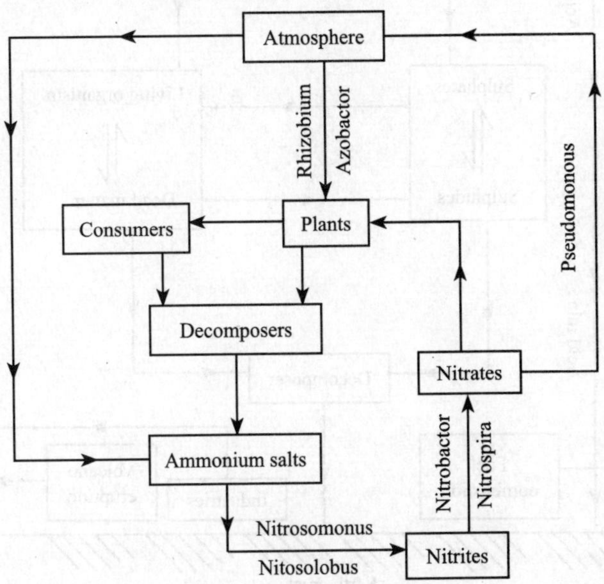

Fig. 1.6 Nitrogen cycle.

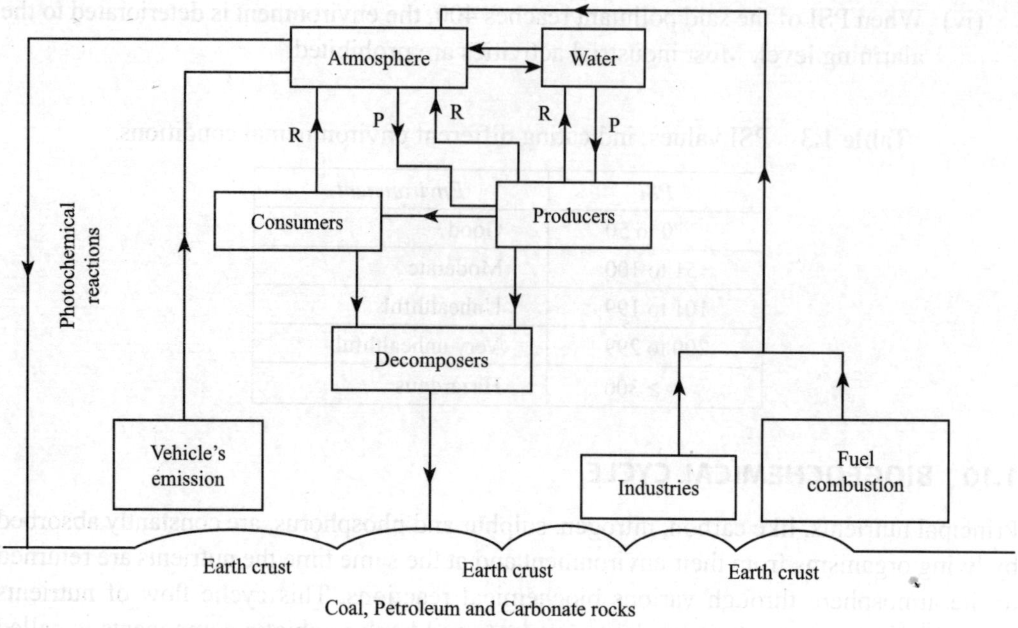

Fig. 1.7 Carbon dioxide cycle.

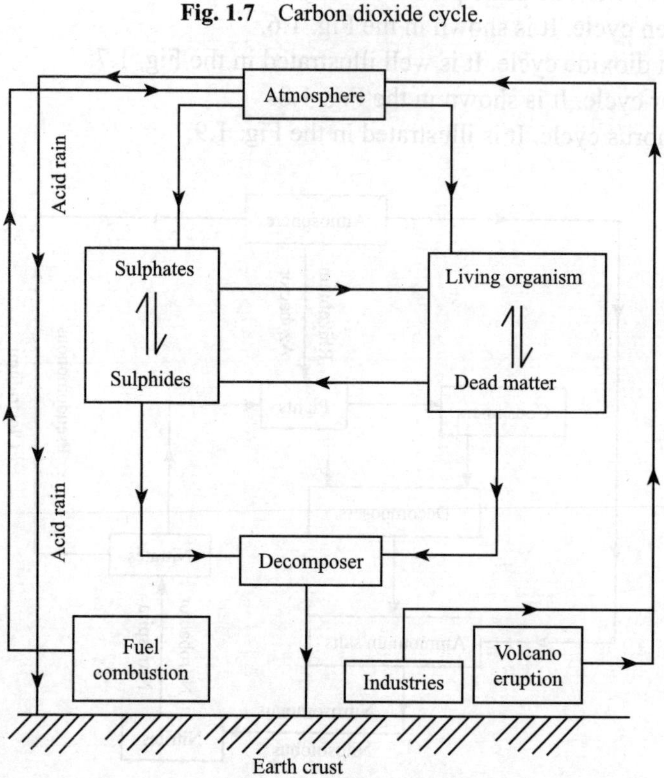

Fig. 1.8 Sulphur cycle.

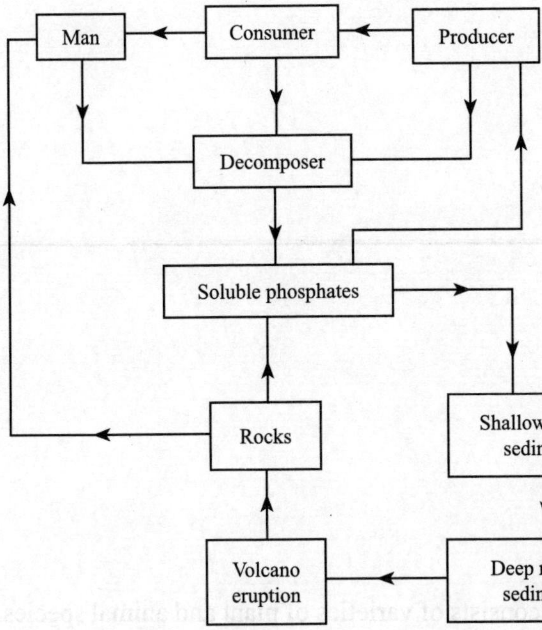

Fig. 1.9 Phosphorus cycle.

2

Biodiversity

2.1 INTRODUCTION

2.1.1 Definition

Each and every ecosystem consists of varieties of plant and animal species. Environmental condition of the Earth has been changing continuously since the formation of the Earth, as a result of which, many species became extinct and some other species have appeared. This natural diversion of species is known as biodiversity.

2.1.2 Classification

Biodiversity can broadly be classified in three categories

(a) **Species diversity** There is a variety of living species or organisms existing in marine, terrestrial and aquatic sources. This is called species diversity.

(b) **Genetic diversity** Variability of genes is available in a single species. This is called genetic diversity.

(c) **Ecological diversity** There is a variety of ecosystems on the planet Earth. This is called ecological diversity. Examples: deserts, forests, ponds, lakes etc.

2.2 IMPORTANCE OF BIODIVERSITY

(a) Different biogeochemical cycles, recycling, purification of ecosystem are well maintained through biodiversity.

(b) Each and every species plays an important role to maintain ecological balance. Thus absence of any species creates ecological imbalance.

(c) Existence of one species depends on existence of others. So species diversity is a must to maintain ecological balance.

(d) Biodiversity boosts ecosystem productivity where each species, no matter how small it is, has an important role to play. For example, a larger number of plant species means a greater variety of crops, greater species diversity ensures natural sustainability for all life forms and healthy ecosystems can better withstand and recover from a different types of natural disasters.

2.3 ENDEMIC SPECIES

An 'Endemic Species' is one that is only found in a particular region and nowhere else in the world. Since these species are not widespread and may be confined to only one or two protected areas, they are of great conservation concern. Some examples of endemic species of India are given below

1. Grey-headed Bulbul.
2. Nilgiri flycatcher.

2.4 BIODIVERSITY HOTSPOTS

Biodiversity hotspots are largely intact areas, supporting natural ecosystems, where native species and communities associated with these ecosystems are well represented. They are also areas with a high diversity of locally endemic species, which are not found or are rarely found outside the hotspot. In India Eastern Himalaya is an example of biodiversity hotspot. Eastern Himalaya is situated between Central Nepal in the west to Myanmar in the east, occupying southeast Tibet in China, Sikkim, North Bengal, Bhutan and North-East India. The area has been declared a biodiversity hotspot by Conservation International.

2.5 MAJOR THREATS TO BIODIVERSITY

1. *Habitat Loss* Human encroachment in the wild areas has resulted in a wide scale destruction of the natural habitat for these wild animals, thus leaving them without a place to live.
2. *Over Hunting* Yet another human threat to biodiversity, over hunting has brought several animal species to the brink of extinction.
3. *Invasive Species* The invasion of non-native species in particular area has also resulted in extinction of species endemic to that region.
4. *Pollution* While the contamination of water bodies can result in extinction of animal species, the contamination of soil can lead to extinction of the plants. Species with limited geographical range are the most affected when it comes to extinction due to pollution.
5. *Climate Change* Over the last few decades, the global surface temperature has experienced an average increase of 1.8 degree Celsius. This has changed the equations in various regions, and affected the plant and animal species, native to this region.

2.6 CAUSES OF LOSS OF BIODIVERSITY

(a) Rainforests are the richest biodiversity. Indiscriminate destruction of these rainforests, particularly in southern hemisphere is the main reason for the loss of biodiversity. Rainforests are destroyed at the rate of 10000 km^2/year. Around 18×10^6 hectares of Amazonian rainforests were converted to agricultural land for the cultivation of coffee. In Germany 2000 hectares of rainforests is destroyed every year for timber production.
(b) Conservation policy of rich countries of the northern hemisphere.
(c) Transfer of environmentally unfriendly technological know how to poor countries of southern hemisphere.
(d) 90% of world's biological resources is confined in the southern hemisphere but all these resources are being transferred to rich countries of the northern hemisphere.
(e) Excessive harvestation of plant and animal resources.
(f) Indiscriminate use of pesticides and germicides in agricultural land.
(g) Different wetlands are either filled up or used as sewerage.

2.7 CONSERVATION OF BIODIVERSITY

Conservation is the protection, preservation, management, or restoration of wildlife and natural resources such as forests and water. Through the conservation of biodiversity the survival of many species and habitats which are threatened due to human activities can be ensured.

2.7.1 Steps to conserve biodiversity and biological resources

(a) Regular survey and assessing the biological resources of each country of the world.
(b) Green technology must be developed and adopted to create sustainable environment.
(c) Quality of biodiversity should be improved by discovering new species and innovating methods for its utilisation.
(d) Conservation of natural resources for sustainable use through alternative planning.
(e) An international agreement was made for conservation of global biodiversity in the name of Intellectual Property Right (IPR), which consists of the following points
    (i) Creation of International Monetary Fund or IMF to uplift underdeveloped and developing countries.
    (ii) Roles of different nations are specified to conserve biodiversity and biological resources.
    (iii) Sharing of biotechnology is another important aspect of IPR. In rich countries of the northern hemisphere several medicines and other special materials are produced from biological resources, which are procured and transferred from poor countries of the southern hemisphere. In return full technological

know-how of the said medicines and special materials should be transferred to those poor countries for their development. USA did not agree with this term of agreement and formulated some new principles, known as General Agreement on Tariffs and Trades (GATT) in 1991.

2.7.2 Conservation status rank

Conservation status is assessed and documented in three distinct geographic scales, (i) global (G), (ii) national (N) and (iii) state/province (S). It is based on a one-to-five scale, ranging from critically imperilled (G1) to demonstrably secure (G5).

The numbers have the following meaning

- 1 = critically imperilled
- 2 = imperilled
- 3 = vulnerable
- 4 = apparently secure
- 5 = secure

2.8 ECOSYSTEM DYNAMICS

Ecosystem dynamics refers to those ecological functions through which an ecosystem becomes self-regulating, self-sustaining, and capable of recovery of external forces, e.g., flood, drought, storm etc. These ecological functions are responsible for continual change in biotic composition and structure in a given ecosystem. These continual changes are known as internal flux of the ecosystem dynamics.

2.9 DIVERSITY INDEX

2.9.1 Definition

A diversity index is a statistical number. It is used to measure the number and varieties of sets available in a given system. Diversity indices can be used in different fields of study to assess the diversity. For example

- In ecology it is used to measure biodiversity in an ecosystem.
- In demography it is used to measure the distribution of population of various demographic groups.
- In economics it is used to measure the distribution over sectors of economic activity in a region.
- In information science it is used to describe the complexity of a set of information.

2.9.2 Biological diversity

Biological diversity refers to the great variety of life. It can be quantified in many different ways. Two important parameters are used to measure the biological diversity, namely,

richness and evenness. Richness is a measure of the number of different kinds of species or organisms present in a particular ecosystem. For example, species richness is the number of different species present in a given ecosystem. On the other hand, evenness compares the population size of each of species present in that ecosystem.

2.9.2.1 Richness factor

Richness of a given ecosystem measures the number of species per sample present in that ecosystem. The higher the number of species present in a sample, the 'richer' the sample is. It does not consider the number of individuals present in each species. It gives as much weight to those species which have very few individuals as to those which have many individuals. In a set of flower, one daisy has similar influence on the richness of an area as 1000 buttercups have.

2.9.2.2 Evenness factor

Evenness is a measure of the relative abundance of the different species present in a given ecosystem.

Let us consider two different samples, each having three different types of wildflowers, namely, daisy, dandelion and buttercup. Details of the samples are tabulated below.

Table 2.1 Examples of diversity of different flower species.

Species	*Number of individuals*	
Flower species	*Sample 1*	*Sample 2*
Daisy	420	50
Dandelion	450	150
Buttercup	530	1200
Total	1400	1400

Both samples have the same richness (3 species) and the same total number of individuals (1400). In the 1st sample, individuals in three species are more or less evenly distributed but in the 2nd sample, buttercup possesses maximum number of individuals with only a few number of daisies and dandelions. Therefore, the 2nd sample is considered to be less diverse than the 1st sample.

Conclusion A community dominated by one or two species is considered to be less diverse than one in which several different species have a similar abundance.

As species richness and evenness increase, diversity increases. Simpson's Diversity Index is a measure of diversity which takes both richness and evenness into account.

2.10 SIMPSON'S DIVERSITY INDICES

The term 'Simpson's Diversity Index' can actually refer to any one of three closely related indices.

2.10.1 Simpson's Index (*D*)

It measures the probability that two individuals randomly selected from a sample will belong to the same species (or some category other than species). There are two versions of the formula for calculating ***D***.

$$1.\ D = \sum\left(\frac{n}{N}\right)^2, \qquad 2.\ D = \frac{\sum n(n-1)}{N(N-1)}$$

Where, n = the total number of organisms of a particular species.
N = the total number of organisms of all species.

The value of D ranges between 0 and 1.

With this index, 0 represents infinite diversity and 1 represents no diversity. That is, the higher the value of D, the lower the diversity. This is neither intuitive nor logical, so to get over this problem, D is often subtracted from 1 to formulate new diversity index.

2.10.2 Simpson's Index of Diversity (1 – *D*)

The value of this index also ranges between 0 and 1, but now, the greater the value, the greater the sample diversity. This makes more sense. In this case, the index represents the probability that two individuals randomly selected from a sample will belong to different species.

Another way of overcoming the problem of the counter-intuitive nature of Simpson's Index is to take the reciprocal of the Index.

2.10.3 Simpson's Reciprocal Index (1/*D*)

The value of this index starts with 1 as the lowest possible figure. This figure would represent a community containing only one species. The higher the value, the greater the diversity is. The maximum value is the number of species (or other category being used) in the sample. For example if there are five species in the sample, then the maximum value is 5.

As an example, let us work out the value of ***D*** for a single quadrat sample of ground vegetation in a woodland. Of course, sampling only one quadrat would not give a reliable estimate of the diversity of the ground flora in the wood. Several samples would have to be taken and analyzed to give a better estimate of overall diversity.

Table 2.2 Calculation of diversity index of different species.

Species	*Number (n)*	*n(n – 1)*
Woodrush	2	2
Holly (seedlings)	8	56
Bramble	1	0
Yorkshire Fog	1	0
Sedge	3	6
Total	15	64

Putting the figures into the formula for Simpson's Index

$$D = \frac{\sum n(n-1)}{N(N-1)} = \frac{64}{15 \times 14} = \frac{64}{210} = 0.305$$

- Simpson's Index = $D = 0.305$.
- Simpson's Index of Diversity = $1 - D = 1 - 0.305 = 0.695$.
- Simpson's Reciprocal Index = $1/D = 1/0.305 = 3.279$

These 3 different values represent the same biodiversity. It is therefore important to ascertain which index has actually been used in any comparative studies of diversity. A value of Simpson's Index of 0.7 is not the same as a value of 0.7 for Simpson's Index of Diversity.

Simpson's Index gives more weight to the more abundant species in a sample. The addition of rare species to a sample causes only small changes in the value of D.

2.11 SHANNON DIVERSITY INDEX

$$H = -\sum_{i=1}^{S} (p_i \ln p_i) = -\sum_{i=1}^{S} \frac{n_i}{N} \ln \frac{n_i}{N}$$

n_i = The number of individuals in species i.
p_i = The relative abundance of each species.
S = The number of species, also called species richness.
N = The total number of all individuals.

The value of H becomes maximum when all species are present in equal numbers and $H_{max} = \ln S$.

Proof

Now, divide the species into two groups. Within each group, the population is evenly distributed among the species present.

- k = The number of individuals in the second group.
- p = The number of species in the second group.
- $n_{i2} = k/p$ = Number of individuals in each species in the second group.
- $(N - k)$ = The number of individuals in the first group.
- $(S - p)$ = The number of species in the first group.
- $n_{i1} = (N - k)/(S - p)$ = Number of individuals in each species in the first group.

$$H = -\sum_{i=1}^{S} (p_i \ln p_i) = -\sum_{i=1}^{S} \frac{n_i}{N} \ln \frac{n_i}{N}$$

$$NH = -\sum_{i=1}^{S} n_i (\ln n_i - \ln N) = -\sum_{i=1}^{S} n_i \ln n_i + \ln N \sum_{i=1}^{S} n_i$$

$$NH = -\sum_{i=1}^{S} n_i \ln n_i + N \ln N$$

$$H_S = NH - N\ln N = -\sum_{i=1}^{S} n_i \ln n_i$$

$$H_S = -\sum_{i=1}^{S-p}\left(\frac{N-k}{S-p}\right)\ln\frac{N-k}{S-p} - \sum_{i=1}^{p}\frac{k}{p}\ln\frac{k}{p} = -(N-k)\ln\frac{N-k}{S-p} - k\ln\frac{k}{p}$$

$$H_S = -(N-k)\ln\frac{N-k}{S-p} - k\ln\frac{k}{p}$$

At the maximum

$$\frac{dH_S}{dk} = 0$$

Thus at the maximum we can write that

$$\ln\frac{N-k}{S-p} + (N-k)\times\frac{1}{(N-k)} - \ln\frac{k}{p} - k\times\frac{1}{k} = 0$$

Or, $\ln\frac{N-k}{S-p} - \ln\frac{k}{p} = 0.$ Or, $\ln\frac{N-k}{S-p} = \ln\frac{k}{p}$

Or, $\frac{N-k}{S-p} = \frac{k}{p}.$ Or, $\frac{k}{p} = \frac{N}{S}$ and $\frac{N-k}{S-p} = \frac{N}{S}$

$n_{i1} = \frac{N-k}{S-p} = \frac{N}{S}$ and $n_{i2} = \frac{k}{p} = \frac{N}{S}$

Or, $n_{i1} = n_{i2} = \frac{N}{S}$ and $p_i = \frac{n_i}{N} = \frac{N/S}{N} = \frac{1}{S}$

$$H_{max} = -\sum_{i=1}^{S}(p_i \ln p_i) = -\sum_{i=1}^{S}\left(\frac{1}{S}\ln\frac{1}{S}\right) = +\sum_{i=1}^{S}\left(\frac{1}{S}\ln S\right) = \ln S$$

3

Atmosphere

3.1 INTRODUCTION

3.1.1 Definition

The atmosphere is a layer of gases, surrounding the Earth. This layer of gases is held by the gravity of the Earth. The atmosphere plays an important role to protect the life on Earth. The atmosphere is consisted of several layers, as below

(i) Troposphere, (ii) Stratosphere, (iii) Mesosphere,
(iv) Thermosphere, (v) Exosphere.

These layers are shown in the Fig. 3.1.

Fig. 3.1 Different levels of atmosphere of Earth.

The first layer, i.e., ***troposphere*** exists up to 11 km above the surface of the Earth.

The second layer, i.e., ***stratosphere*** starts above troposphere and exists up to 50 km above the surface of the Earth. There is a distinct layer, separating troposphere and stratosphere. This layer is called tropopause.

The next layer starts above stratosphere and exists up to 85 km above the surface of the Earth. This layer is called ***mesosphere***. There is a distinct layer, separating stratosphere and mesosphere. This layer is called stratopause.

The next layer, called ***thermosphere***, starts above mesosphere and exists up to 690 km above the surface of the Earth. There is a distinct layer, separating mesosphere and thermosphere. This layer is called mesopause.

The next layer, called ***exosphere***, starts above thermosphere and exists up to 10000 km above the surface of the Earth.

The main constituents of thermosphere and exosphere are ions and hence these two layers are collectively called ***ionosphere***. Now-a-days mesosphere is also included in the ionosphere. The atmosphere above 10 km is called outer space. The Kármán line at 100 km is considered the boundary between atmosphere and outer space.

3.2 PHYSICAL PROPERTIES

Mass, density and pressure

The atmosphere contains huge amount of water vapour. The mean total mass, including the water vapour, of the atmosphere is 5.148×10^{18} kg. 3/4th of this mass is confined in troposphere. The mean mass of water vapour is approximately 1.27×10^{16} kg. The mean dry air mass is 5.135×10^{18} kg. 50% of this air mass is concentrated below 5.6 km above the sea level. 90% of atmospheric mass is concentrated below 16 km above the sea level.

The air is assumed to behave as an ideal gas. The average molecular weight of air is taken as 29 g/mol. Using ideal gas equation density of air can be calculated and its value is around 0.0012 gm/ml or 1.2 gm/l. This density almost remains constant up to 8.5 km above the sea level.

Atmospheric pressure is assumed to be 1 atm at sea level. However, the pressure decreases with height.

3.2.1 Troposphere

The main constituents of troposphere are nitrogen, oxygen, argon, carbon dioxide and water vapour. Besides that ozone, ammonia, nitrogen oxides, hydrogen and other inert gases are also present in trace amount.

Table 3.1 Principal components of troposphere.

Constituent	*% by volume*
Nitrogen	78%
Oxygen	20.9%
Argon	0.93%
Carbon dioxide	0.039% or 390 *ppm*
Water vapour	Approx. 0.4%

At sea level the pressure in open atmosphere is universally accepted as 1 atm. However, pressure decreases as we move upward and at the same time temperature of troposphere also decreases with altitude. Temperature near the Earth surface is around +15°C and near tropopause, it is around –56°C. This change in temperature with altitude is known as lapse rate. So lapse rate is negative in troposphere. Ozone is present in trace amount in troposphere. The ratio of concentration of oxygen to that of ozone is of the order of 10^{25}.

3.2.2 Stratosphere

This region starts above tropopause and exists up to 50 km above the surface of the Earth. The main constituents of stratosphere are oxygen and ozone. However, unlike troposphere, ozone concentration is very high in this region. The ratio of concentration of oxygen to that of ozone is of the order of 10^5. The thick ozone layer absorbs harmful UV rays, protecting life on Earth. Due to this absorption of UV rays temperature of this region increases with altitude and hence lapse rate is positive. Temperature near stratopause is around +2°C.

3.2.3 Mesosphere

This region exists above stratosphere and extends up to 85 km above the surface of the Earth. The main constituents are O_2, N_2 and NO. It is also called D-layer of ionosphere. Lapse rate is negative. Temperature decreases from +2°C to –85°C. The transition region, separating mesosphere and thermosphere is called mesopause.

3.2.4 Thermosphere

This region exists above the mesopause and extends up to 690 km above the surface of the Earth. The main constituents are ions, namely, O^+, O_2^+, N_2^+ etc. Thermosphere is divided into two ionospheric layers, namely, E-layer and F-layer. Lapse rate is positive and temperature increases steadily from –85°C to +1500°C. The transition region, separating thermosphere and exosphere, is called thermopause.

3.2.5 Exosphere

The outermost layer of the Earth atmosphere is called exosphere. It exists up to 10000 km above the Earth surface. This layer is mainly composed of hydrogen and helium. The particles in this region are wide apart so that collision rarely occurs even within hundreds of kilometres. As the frequency of collision is very less, atmosphere does not behave like a fluid in this region. Temperature of this region is nearly constant around +1500ºC.

3.2.6 Ionosphere

Temperature is very high in thermosphere and exosphere, as a result of which, the atmosphere is mainly composed of ions, like O^+, O_2^+, N_2^+ etc. These ions play an important role in generating atmospheric electricity and magnetism. These ions also influence the propagation of radio waves. That is why, thermosphere and exosphere is collectively called ionosphere. Thermosphere constitutes E-layer and exosphere constitutes F-layer in the ionosphere. Mesosphere does not contain significant number of ions at night but in day time in presence of solar radiation sufficient number of ions are generated in mesosphere. Then mesosphere is also included in the ionosphere and is called D-layer.

D-layer

During the day time hard X-rays, having wavelength < 1 nm, penetrates into D-layer and ionize the atmosphere of mesosphere. However, ions recombine at a very fast rate and hence net ionizing effect is very low. Only ions are responsible for reflecting radio waves. As the number of ions is very less, most of the high frequency (HF) radio waves are not reflected by D-layer. HF radio waves with frequency at 10 MHz or below are absorbed by D-layer.

E-layer

This layer is spread over 90 to 120 km above the surface of the Earth. Ionization mainly occurs due to soft X-rays (wavelength range 1–10 nm) and far UV-rays. The main constituent of this layer is O_2^+. In day time ionization rate is much higher than recombination rate. These ions reflect radio waves of frequency <<10 MHz. In some region ionization rate is so fast that huge number of ions are generated. That layer can reflect radio waves with frequency 25–225 MHz. That region is called sporadic E-layer. However, at night recombination rate is high and hence E-layer shortens and moves up.

F-layer

This is the topmost layer of the ionosphere. It is also known as Appleton layer, ranging from 150 km to 800 km above the surface of the Earth. Very hard Ultraviolet rays (10–100 nm) ionizes atomic oxygen to form O^+ and this is the main constituent of this layer. At night F-layer exists as one layer but in daytime it slits into two layers F_1 and F_2. These F-layers are the most effective to reflect and propagate radio waves or signals over long distances on the surface of the Earth.

3.2.7 Meteorology

Study on the dynamics of the atmosphere and air quality is called meteorology. Air quality depends on several factors

(a) Wind speed.
(b) Direction of blowing wind.
(c) Temperature profile.
(d) Availability of solar energy and duration of sunlight to execute different photochemical reactions.

Weather It is the short term variations of atmosphere in terms of temperature, pressure, wind, moisture, cloudiness, precipitation and visibility.

Climate It represents the average weather of an area over a long period of time.

3.3 LAPSE RATE

Temperature of atmosphere changes continuously with altitude. The rate of change of temperature with altitude is known as lapse rate.

Lapse rate may be classified into two different types

(a) Adiabatic lapse rate or ALR, represented by Γ_{ALR}.
(b) Environmental lapse rate or ELR, represented by Γ_{ELR}.

3.3.1 Adiabatic lapse rate or ALR

It is a theoretical concept. Consider a parcel of air containing pollutants is moving vertically such that no heat transfer occurs through parcel wall. So adiabatic condition is maintained within the parcel or $dQ = 0$. As the air parcel is moving upward, temperature of the parcel changes with altitude. Under this condition the rate of change of temperature with altitude is known as ALR. It is represented as Γ_{ALR}.

3.3.2 Environmental lapse rate or ELR

As the air parcel is moving upward, temperature of the parcel changes with altitude and simultaneously temperature of the surrounding air changes with altitude. This rate of change of temperature of surrounding air with altitude is known as environmental lapse rate or ELR. It is represented as Γ_{ELR}. The magnitude of ELR depends on several factors – (i) Altitude, (ii) Season. The magnitude of ELR is lower in winter season or rainy season. (iii) Type of land. The magnitude of ELR is lower in plateau areas than that in coastal areas. (iv) Particulate matter content.

3.3.3 Calculation of Dry Adiabatic Lapse Rate or DALR

Similar to the concept of light, air is considered as numerous mass packets like photons. Each and every mass packet is surrounded by an adiabatic wall as shown in the Fig. 3.2.

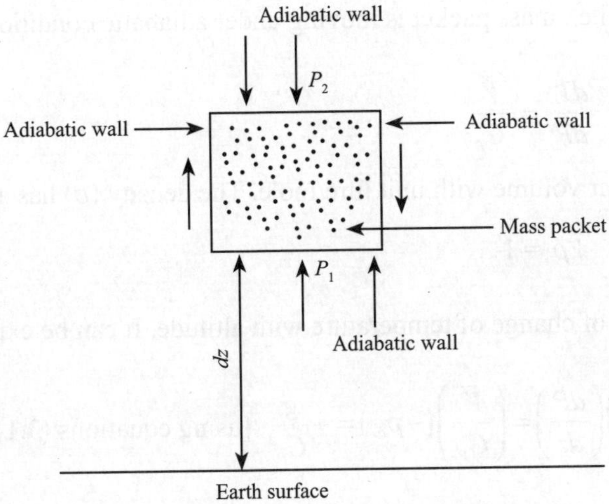

Fig. 3.2 Force balance on mass packet.

ρ = Density of a mass packet in mole/m^3.
$\bar{V}$ = Molar volume of the mass packet.
dz = Height of the mass packet from the surface of the Earth.
A = Area of the mass packet.

Hence, volume of the mass packet = $V = n\bar{V}$, n = No. of moles of dry air.

P_1 = Pressure at the bottom surface of the mass packet, arises due to weight of the air acting on the mass packet and weight of the mass packet.

P_2 = Pressure at the bottom surface of the mass packet, arises due to weight of the air acting on the mass packet.

Potential energy of the mass packet is $mgdZ = (\rho \times V)\, gdZ$ Joule.

So, pressure at the bottom surface due to potential energy is $mgdZ/V = \rho gdZ\ J/m^3$ or Pa.

Thus at equilibrium, pressure at the bottom surface is equal to pressure at the top surface. So, we can write

So, $$P_1 = P_2 + \rho gdz$$

$\therefore$ Change of pressure, $dP = P_2 - P_1 = -\rho gdz$

Or, $$\frac{dP}{dz} = -\rho g \tag{3.1}$$

The negative sign indicates that pressure decreases with altitude.

Again for 1 mole of an ideal gas the following thermodynamic relation holds good

$$dH = dQ + \bar{V}dP \text{ and } dH = C_P\, dT$$

Or, $$dQ = C_P\, dT - \bar{V}dP$$

As the system, i.e., mass packet is moving under adiabatic condition, we have $dQ = 0$

Hence, $$\frac{dT}{dP} = \frac{\bar{V}}{C_P} \tag{3.2}$$

where, $\bar{V}$ is the molar volume with unit litre/mole. The density (ρ) has a unit of mole/litre.

Thus, $$\bar{V}\rho = 1 \tag{3.3}$$

$\frac{dT}{dz}$ represents rate of change of temperature with altitude. It can be expressed as

$$\frac{dT}{dz} = \left(\frac{dT}{dP}\right)\left(\frac{dP}{dz}\right) = \left(\frac{\bar{V}}{C_P}\right)(-\rho g) = -\frac{g}{C_P} \quad \text{[using equations (3.1), (3.2) and (3.3)]}$$

Or, $$\frac{dT}{dz} = -\frac{g}{C_P} \tag{3.4}$$

The negative sign indicates that temperature decreases with altitude.

The value of g is 9.81 m/s^2 and C_p for dry air is 1005 J/(kg.°C). Putting the values of g and C_P in the equation (3.4), we get

$$\frac{dT}{dz} = -\frac{g}{C_P} = -\frac{9.81}{1005}\text{°C/m} = -0.00976\text{°C/m} = -9.76\text{°C/km}$$

Or, $$\Gamma_{ad} = \frac{dT}{dz} = -9.76\text{°C/km} \tag{3.5}$$

The above value of ALR is obtained when air parcel is very dry with little or no moisture. The C_P value of air is the lowest at this condition and hence magnitude of ALR is maximum. This value of ALR is often called Dry Adiabatic Lapse Rate or DALR (Γ_{ALR}). Temperature of the air parcel is higher than dew point temperature.

If the air parcel is completely saturated with moisture, C_P value of air is very high and hence magnitude of ALR is minimum, at this condition $dT/dz = -5.4$°C/km. Temperature of the air parcel is lower than dew point temperature. This value of ALR is often called Saturated Adiabatic Lapse Rate or SALR.

In actual practice, the magnitude of ALR lies within 5.4°C/km and 9.76°C/km. Average value of ALR is –6.5°C/km.

The lapse rate of different regions of atmosphere is shown in the Fig. 3.3.

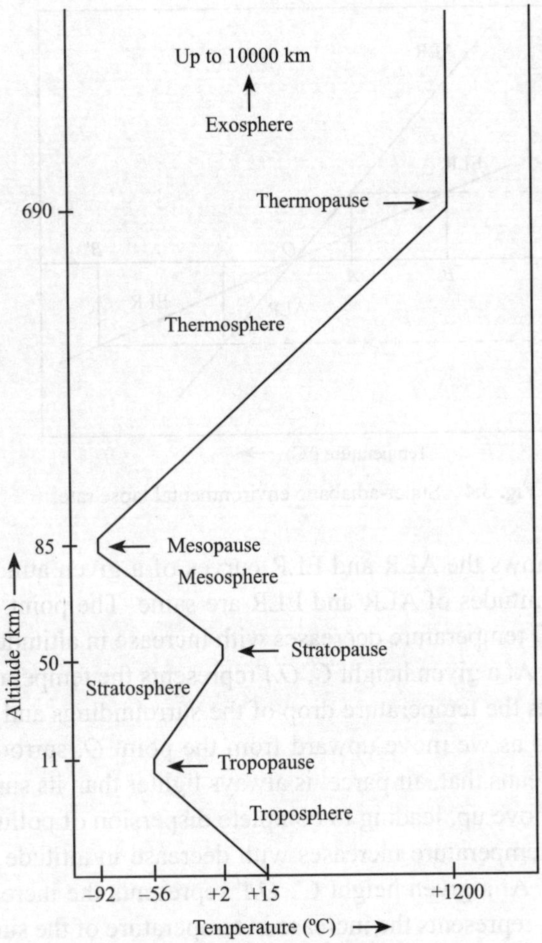

Fig. 3.3 Lapse rate at different regions of Earth's atmosphere.

3.4 ATMOSPHERIC STABILITY

The stability of a given part of atmosphere depends on the magnitudes of ALR and ELR.

3.4.1 Unstable atmosphere

There is one atmospheric condition under which the air parcel can move upwards or move downwards continuously without any hindrance, i.e., rapid vertical mixing takes place.

The atmosphere is said to be unstable and the corresponding ELR is called super-adiabatic environmental lapse rate. It is well illustrated in the Fig. 3.4.

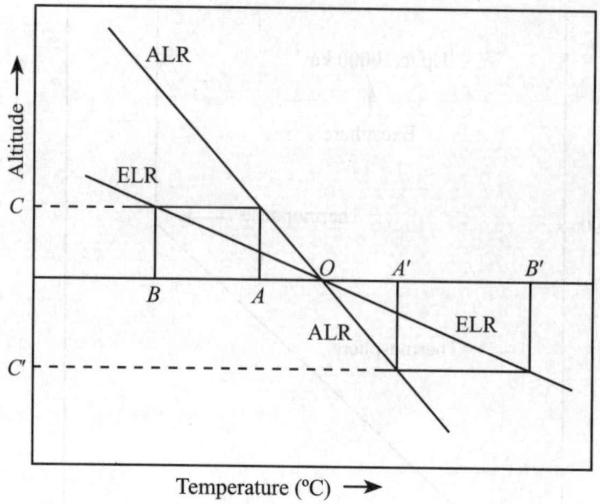

Fig. 3.4 Super-adiabatic environmental lapse rate.

The above figure shows the ALR and ELR curves of a given atmospheric condition. At the point *O* the magnitudes of ALR and ELR are same. The point *O* is called neutral point. Above the point *O* temperature decreases with increase in altitude and ALR curve is steeper than ELR curve. At a given height *C*, *OA* represents the temperature drop of the air parcel and *OB* represents the temperature drop of the surroundings and *OB* is greater than *OA*. So it is evident that as we move upward from the point *O*, surroundings cool faster than air parcel, which means that, air parcel is always lighter than its surroundings. So, the air parcel continues to move up, leading to complete dispersion of pollutants.

Below the point *O* temperature increases with decrease in altitude and ALR curve is steeper than ELR curve. At a given height *C′*, *OA′* represents the increase in temperature of the air parcel and *OB′* represents the increase in temperature of the surroundings. Again, *OB′* is greater than *OA′*. So it is evident that as we move downward from the point *O*, surroundings is heated up faster than air parcel, which means that, air parcel is always heavier than its surroundings. So, the air parcel continues to move down, leading to complete dispersion of pollutants.

At this condition the atmosphere is called unstable and corresponding ELR is called super-adiabatic ELR.

3.4.2 Stable atmosphere

There is one atmospheric condition under which the air parcel can move upwards up to a certain height, where the density of air parcel is same as that of surroundings. Above that height air parcel becomes heavier than its surroundings and it starts coming down. Similarly the parcel can move downwards up to a certain limit, at which the density of air parcel is same as that of surroundings. Below that height air parcel becomes lighter than its surroundings and then starts moving up. So the movement of air parcel is restricted within a certain region, i.e., no vertical mixing takes place. The atmosphere is said to be stable and the corresponding ELR is called sub-adiabatic environmental lapse rate. It is well illustrated in the Fig. 3.5.

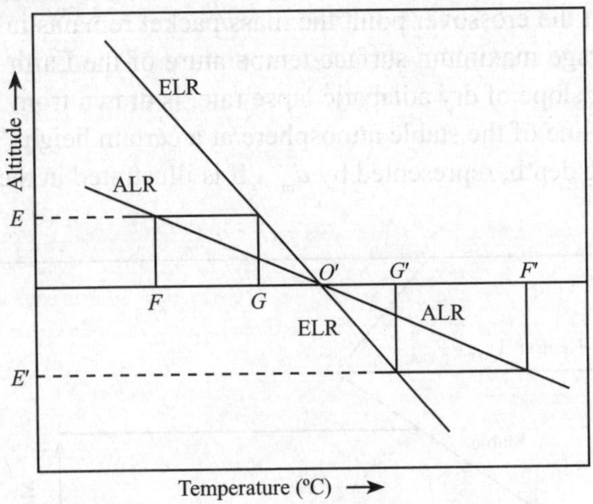

Fig. 3.5 Sub-adiabatic environmental lapse rate.

The above figure shows the ALR and ELR curves of a given atmospheric condition. At the point *O′* the magnitudes of ALR and ELR are same. The point *O′* is called neutral point. Above the point *O′* temperature decreases with increase in altitude and ELR curve is steeper than ALR curve. At a given height *E*, *O′F* represents the temperature drop of the air parcel and *O′G* represents the temperature drop of the surroundings. Again, *O′F* is greater than *O′G*. So it is evident that as we move upward from the point *O′*, air parcel cools faster than surroundings. Thus after a certain altitude air parcel becomes heavier than surroundings and it, i.e., air parcel, starts coming down. So, the air parcel is restricted to move up to a certain height, leading to incomplete dispersion of pollutants.

Below the point *O′* temperature increases with decrease in altitude and ELR curve is steeper than ALR curve. At a given height *E′*, *O′F′* represents increase in temperature of the air parcel and *O′G′* represents increase in temperature of the surroundings. Again, *O′F′* is greater than *O′G′*. So it is evident that as we move downward from the point *O′*, air parcel is heated up faster than surroundings. Thus after a certain altitude air parcel becomes lighter than its surroundings and it, i.e., air parcel, starts moving up. So, downward movement of the air parcel is also restricted to a certain height, leading to incomplete dispersion of pollutants.

At this condition the atmosphere is called stable and corresponding ELR is called sub-adiabatic ELR.

3.5 MAXIMUM MIXING DEPTH

Temperature profile determines atmospheric stability, which in turn governs the vertical dispersion of pollutants. If the temperature of mass packet is higher than that of surroundings, mass packet continues to rise and the same continues to fall if the mass packet is colder than the surroundings. At the crossover point the mass packet remains in equilibrium position. T_{max} represents average maximum surface temperature of the Earth for a given month. A straight line, having slope of dry adiabatic lapse rate, is drawn from T_{max}. The straight line meets the adiabatic line of the stable atmosphere at a certain height. This height is known as maximum mixing depth, represented by d_{max}. It is illustrated in the Fig. 3.6.

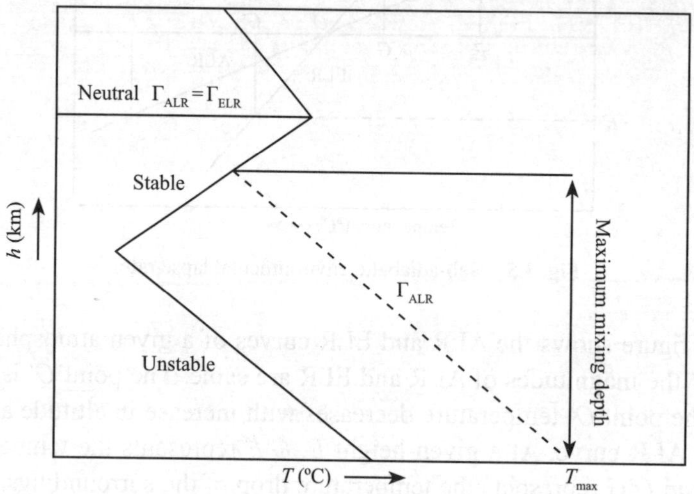

Fig. 3.6 Maximum mixing depth.

3.5.1 Ventilation coefficient

Wind speed generally increases with altitude due to decrease in density of air. Wind speed and altitude are correlated as follows

$$\frac{u_2}{u_1} = \left(\frac{z_2}{z_1}\right)^p$$

u_1 and u_2 are the wind speeds at altitudes z_1 and z_2 respectively. p is a dimensionless fraction, varying with the atmospheric stability.

(a) For neutral atmosphere $p = 0.25$ and $|\Gamma_{ALR}| = |\Gamma_{ELR}|$
(b) For unstable atmosphere $p < 0.25$ and $|\Gamma_{ALR}| > |\Gamma_{ELR}|$
(c) For stable atmosphere $p > 0.25$ and $|\Gamma_{ALR}| < |\Gamma_{ELR}|$

For smooth terrain and flat fields, the value of p should be multiplied by a factor 0.6. The following table (Table 3.2) illustrates the variation of p with the stability of atmosphere.

Table 3.2 Variation of p with atmospheric stability.

Stability class	*Atmospheric stability*	*Value of p*
A	Very unstable	0.10
B	Moderately unstable	0.15
C	Slightly unstable	0.20
D	Neutral	0.25
E	Slightly stable	0.40
F	Stable	0.60

Thus ventilation coefficient (η_{vc}) is expressed as

$$\eta_{vc} = \text{Maximum mixing depth} \times \text{Average wind speed within mixing depth}$$
$$= d_{\max} \times v_{av}$$

The unit of η_{vc} is m²/s.

Significance of η_{vc} It indicates the capability of atmosphere to disperse pollutants. The value of $\eta_{vc} < 6000$ m²/s indicates poor dispersion of air pollutants and hence air pollution potential is high.

3.6 PLUME RISE AND DISPERSION

3.6.1 Introduction

Main source of pollutants are automobile exhausts, chemical factory exhausts and power plant exhausts. These pollutants are then carried away by wind. Pollutants are generally released from stationary point sources, which are known as stacks. Pollutants usually emit from stacks in the form of cloud, known as plume. The plume is then travelling in the atmosphere and during its travelling ambient air is often mixed into the plume. The phenomenon is called entrainment. As the plume density is less than the outdoor air density, buoyancy force is predominant. Furthermore, the plume has a definite velocity and momentum. The combined effect of momentum and buoyancy force causes the plume to rise in the atmosphere. This is called plume rise. However, during travelling plume diameter increases. At a certain height, plume becomes heavier than outdoor air and it starts moving down. The distance between surface of the Earth and the centerline of the plume is called effective stack height, represented as H.

If h_s is the physical stack height and Δh is the plume rise, H can be written as $H = h_s + \Delta h$.

The magnitude of Δh depends on temperature difference of stack gas (T_s) and outdoor air (T_a). Momentum of the emitted gas depends on stack diameter and volumetric flow rate of the gas.

Several research works have been documented on the plume rise model. Out of them most common is Briggs' plume model. The working formula to calculate plume rise is given below

$$\Delta h = \frac{1.6 F^{\frac{1}{3}} x^{\frac{2}{3}}}{\bar{u}},$$

where, $$F = \frac{g}{\pi}\dot{V}\left(\frac{T_s - T_a}{T_s}\right)$$

where, Δh = Plume rise above stack.
F = Buoyancy flux.
x = Downwind distance from the stack.
g = Acceleration due to gravity.
$\dot{V}$ = Volumetric flow rate of the stack gas.
T_s = Temperature of the stack gas.
T_a = Temperature of the outdoor air.

Δh is measured with respect to an imaginary centreline of the plume. The centreline is drawn through those points, where the average concentration of pollutants is maximum. However, it is very difficult to calculate the pollutant concentration at a given point. Gaussian distribution model is widely accepted to calculate pollutant concentration. Gaussian distribution model is discussed in detail in '*Workbook of Atmospheric Dispersion Estimates*' by D.B. Turner.

3.6.2 Gaussian distribution of pollutant

Consider a small control volume, filled with pollutant. $C(x, y, z)$ is the concentration of the pollutant. As time passes, concentration of the pollutant C changes because the pollutant is dispersing in a turbulent flow. Thus mass balance over the control volume can be written as

$$\frac{dC}{dt} + u\frac{dC}{dx} + \frac{d}{dy}\left(K_y \frac{dC}{dy}\right) + \frac{d}{dz}\left(K_z \frac{dC}{dz}\right) + \dot{Q}$$

C = Mean concentration of diffusing substance at a point (kg/m³).
K_y = Eddy diffusivity along y-axis (m²/s).
K_z = Eddy diffusivity along z-axis (m²/s).
u = Mean wind velocity along the x-axis (m/s).
$\dot{Q}$ = Concentration flow rate of the pollutant from the stack [kg/(m³ – s)]

$\left(\frac{dC}{dt} + u\frac{dC}{dx}\right)$ represents the change in concentration of pollutant with time as well as due to advection cloud by the mean wind.

$\frac{d}{dy}\left(K_y \frac{dC}{dy}\right)$ represents turbulent diffusion of the pollutant with respect to the centreline of the plume along horizontal direction.

$\frac{d}{dz}\left(K_z \frac{dC}{dz}\right)$ represents turbulent diffusion of the pollutant with respect to the centreline of the plume along vertical direction.

The following assumptions are made to simplify the above equation

(a) The pollutant concentrations do not affect the flow field. It is generally termed as passive dispersion.

(b) Molecular diffusion and longitudinal diffusion processes are considered negligible.

(c) The flow of pollutant is considered incompressible.

(d) The fluctuations of wind velocity and pollutant concentration do not happen. Only mean values of wind velocity and pollutant concentration are considered and these mean values are calculated within the time span of 10–60 minutes.

(e) The turbulent fluxes are linearly related to the gradients of the mean concentrations. If $\overline{v'c'}$ and $\overline{w'c'}$ represent mean concentrations of the pollutant along horizontal and vertical directions respectively, we can write $\overline{v'c'} = -K_y \frac{dC}{dy}$ and $\overline{w'c'} = -K_z \frac{dC}{dz}$.

(f) The wind speed is steady and uniform.

(g) The plume spread curve is considered to be normal distribution curve.

(h) The emission rate, i.e., $\dot{Q}$ is constant and continuous.

(i) Total reflection of plume takes place at the surface.

The Gaussian distribution curve is shown in the Fig. 3.7.

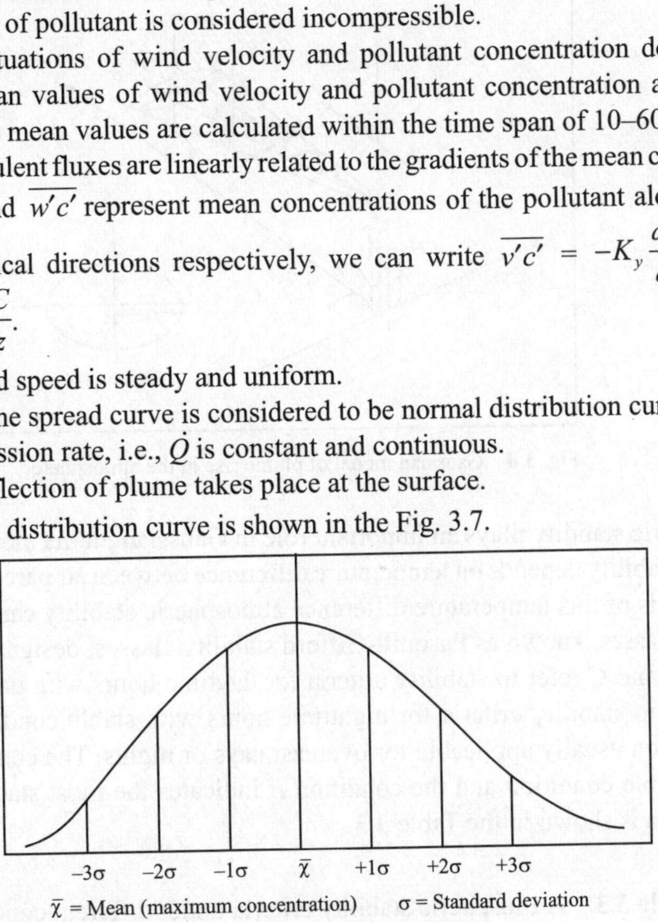

Fig. 3.7 Gaussian distribution curve.

$\overline{\chi} = C =$ Mean concentration. According to the distribution curve, it is also the maximum concentration.

The corresponding mathematical form is given by

$$C = \frac{\dot{Q}}{4\pi x\sqrt{K_y K_z}} \exp\left[\frac{-y^2}{4K_y\left(x/u\right)}\right] \exp\left[\frac{-(z-H)^2}{4K_z\left(x/u\right)}\right] \exp\left[\frac{-(z+H)^2}{4K_z\left(x/u\right)}\right]$$

According to Gaussian dispersion model, plume is dispersed in both horizontal and vertical directions. σ_y and σ_z represent standard deviations from normal on the Gaussian distribution curve in the y- and z-directions. Both σ_y and σ_z depend on wind speed, cloud cover and surface heating by the Sun. According to Gaussian model, the plume rise in the atmosphere is shown in the Fig. 3.8.

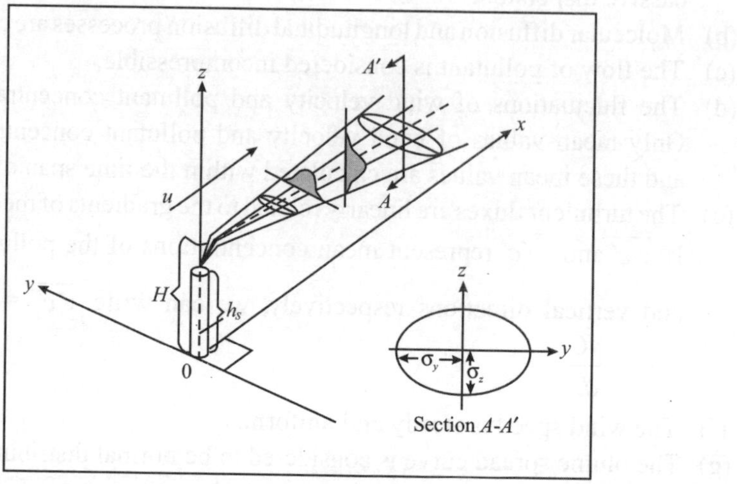

Fig. 3.8 Gaussian model of plume rise in the atmosphere.

Atmospheric stability plays an important role in Gaussian plume model. However, this atmospheric stability depends on temperature difference between air parcel and surrounding air. On the basis of this temperature difference atmospheric stability can be classified into six different classes, known as Pasquill-Gifford stability classes, designated as *A*, *B*, *C*, *D*, *E* and *F*. *A*, *B* and *C* refer to stability criteria for daytime hours with unstable conditions. *E* and *F* refer to stability criteria for nighttime hours with stable conditions. *D* refers to neutral condition usually applicable for overcast days or nights. The condition *A* indicates the most unstable condition and the condition *F* indicates the most stable condition. The stability criteria is shown in the Table 3.3.

Table 3.3 Atmospheric stability criteria under different conditions.

Surface wind speed at a height of 10 m	*Intensity of solar flux in daytime*			*Cloud cover at nighttime*	
	High	*Moderate*	*Low*	*Low*	*High*
< 2 m/s	A	A–B	B	–	–
2 to 3 m/s	A–B	B	C	E	F
3 to 5 m/s	B	B–C	C	D	E
5 to 6 m/s	C	C–D	D	D	D
> 6 m/s	C	D	D	D	D

According to Gaussian model, plume is dispersed in both horizontal and vertical directions. σ_y and σ_z represent horizontal and vertical dispersion coefficients respectively. As the atmospheric condition becomes more and more unstable, i.e., stability criteria moves from *F* to *A*, both σ_y and σ_z increase. However, change in σ_z is more abrupt than that of σ_y as shown in the Fig. 3.9.

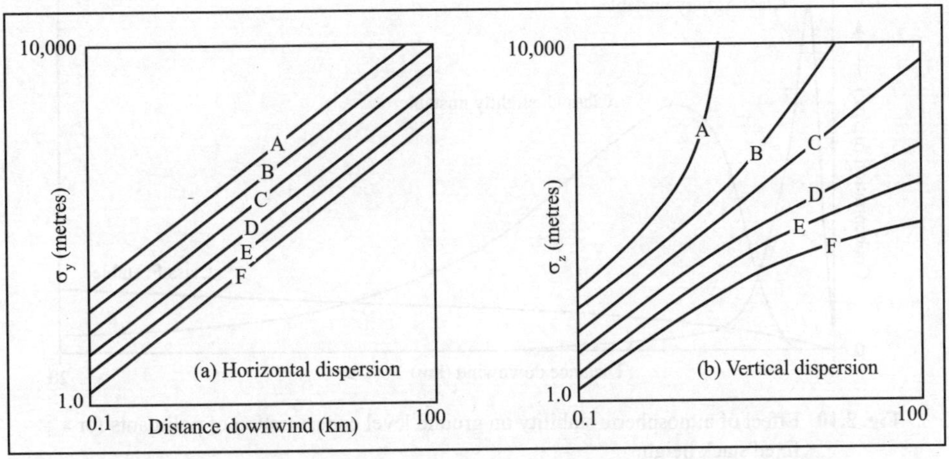

Fig. 3.9 Influence of atmospheric stability on standard deviations (σ_y, σ_z).

Mathematically, σ_y and σ_z can be written as

$$\sigma_y = \sqrt{2K_y\left(\frac{x}{u}\right)} \quad \text{and} \quad \sigma_z = \sqrt{2K_z\left(\frac{x}{u}\right)}$$

Thus the final form of Gaussian plume equation when plume is released at $z = H$ is given by

$$C = \frac{\dot{Q}}{2\pi u\sigma_y\sigma_z}\exp\left[-\frac{y^2}{2\sigma_y^2}\right]\exp\left[-\frac{(z-H)^2}{2\sigma_z^2}\right]\exp\left[-\frac{(z+H)^2}{2\sigma_z^2}\right]$$

Case-1 Consider the distribution of pollutant at the ground level, where people are living. Then $z = 0$. Thus the above equation becomes

$$C = \frac{\dot{Q}}{\pi u\sigma_y\sigma_z}\exp\left[-\frac{y^2}{2\sigma_y^2}\right]\exp\left[-\frac{H^2}{2\sigma_z^2}\right]$$

Case-2 If the source is at ground level, $H = 0$. Thus the above equation becomes

$$C = \frac{\dot{Q}}{\pi u\sigma_y\sigma_z}\exp\left[-\frac{y^2}{2\sigma_y^2}\right]\exp\left[-\frac{z^2}{2\sigma_z^2}\right]$$

Both the stability factor and stack height have profound influence on ground level concentration of pollutants. These are well illustrated in the Fig. 3.10 and Fig. 3.11 respectively.

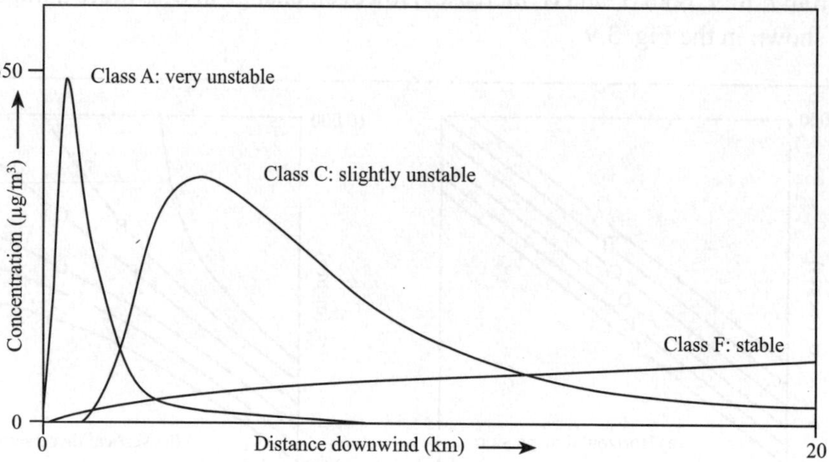

Fig. 3.10 Effect of atmospheric stability on ground level concentration of pollutants for a fixed stack height.

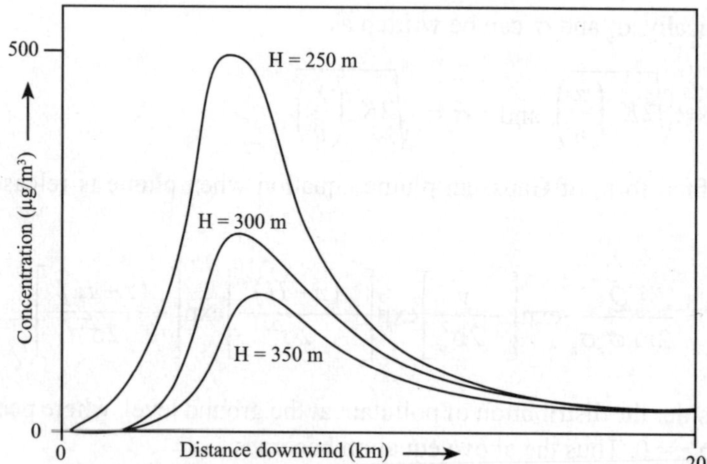

Fig. 3.11 Effect of stack height on ground level concentration of pollutants at a given atmospheric stability condition.

3.6.3 Plume model

After releasing from the stack plume is dispersed in the atmosphere. However the dispersion is highly influenced by wind speed as well as environmental lapse rate. Depending on meteorological conditions plume behaviour can be classified into following categories.

3.6.3.1 Coning

When the smoke is released from the stack to neutrally stable atmosphere, i.e., $\Gamma_{ALR} \approx \Gamma_{ELR}$, the vertical air temperature gradient lies between dry adiabatic and isothermal conditions. The air is slightly unstable so that low degree of some horizontal and vertical mixing takes place. Coning generally occurs during cloudy or windy weather. In this case, plume dispersion is symmetrical in both upward and downward directions. Thus plume rise appears as cone like structure. Hence it is called coning. The corresponding lapse rate diagram and coning phenomenon are shown in the Fig. 3.12.

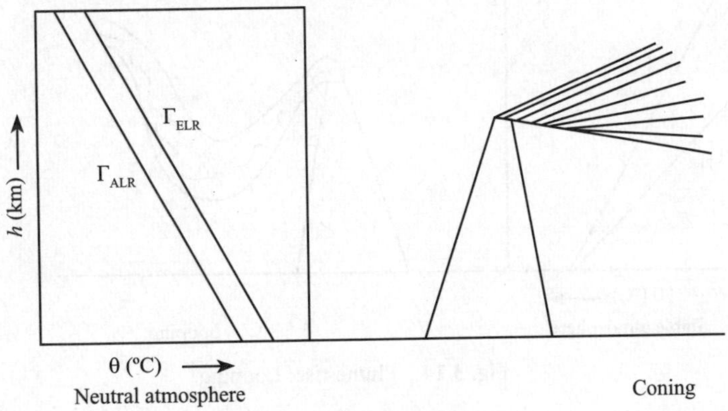

Fig. 3.12 Plume rise: Coning.

3.6.3.2 Fanning

The smoke is released from the stack to stable atmosphere, i.e., Γ_{ALR} is greater than Γ_{ELR}. The air temperature increases with altitude (inversion). Under this condition plume is not allowed to spread vertically although it spreads horizontally. It is called fanning. Fanning plume does not reach ground level unless the inversion is broken. Night with light winds and clear sky favours the propagation of fanning plume. The corresponding lapse rate diagram and fanning phenomenon are shown in the Fig. 3.13.

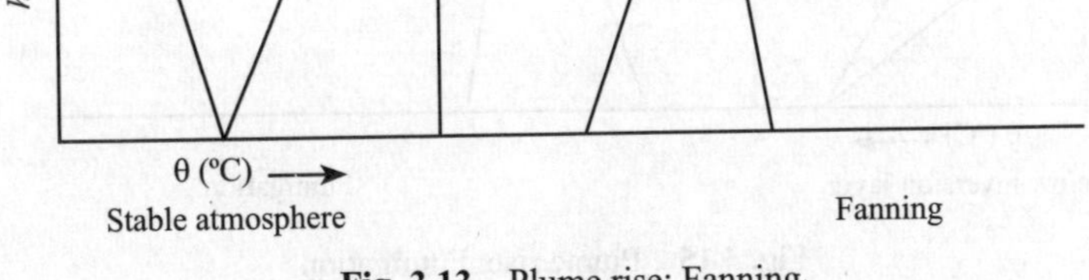

Fig. 3.13 Plume rise: Fanning.

3.6.3.3 Looping

The smoke is released from the stack to unstable atmosphere, i.e., Γ_{ALR} is less than Γ_{ELR}. So the primary condition is super-adiabatic lapse rate and solar heating. Under this condition unstable air forces the plume to spread both in upward and downward directions so that the plume appears as a loop. It is called looping. As the large eddies tend to disperse pollutants over a wide region, high ground level concentration of pollutants is often found close to the stack. The corresponding lapse rate diagram and looping phenomenon are shown in the Fig. 3.14.

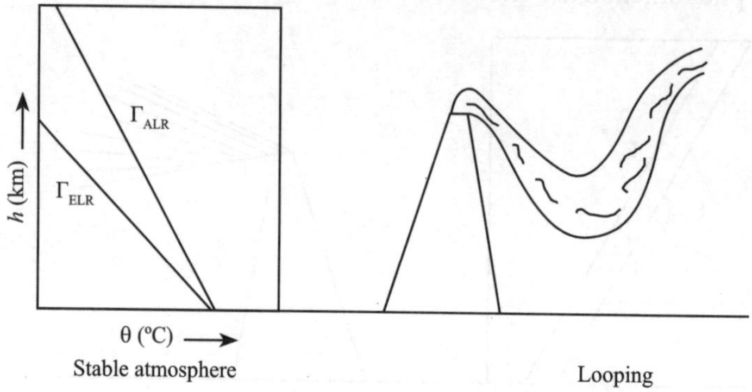

Fig. 3.14 Plume rise: Looping.

3.6.3.4 Fumigation

The atmosphere has inversion layer. Due to solar heating super-adiabatic lapse rate is observed at the ground level, however sub-adiabatic condition prevails above the inversion layer. Clear sky and light wind favours this condition. Stack height is low enough to emit plume below inversion layer. Under this condition movement of pollutants in upward direction is restricted but pollutants move in downward direction, leading to high ground level concentration. This condition is called fumigation. The corresponding lapse rate diagram and fumigation phenomenon are shown in the Fig. 3.15.

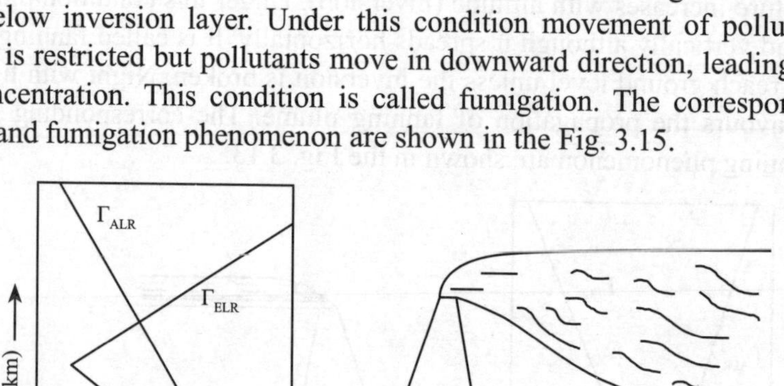

Fig. 3.15 Plume rise: Fumigation.

3.6.3.5 Lofting

In this case also sufficient solar heating is required to develop inversion layer but super-adiabatic layer is developed above the inversion layer. Stack height is high enough to release plume above the inversion layer. So the vertical mixing occurs freely in upward direction but it is restricted in downward direction. It is called lofting. In case of lofting plume pollutants will generally not reach the ground surface. The corresponding lapse rate diagram and lofting phenomenon are shown in the Fig. 3.16.

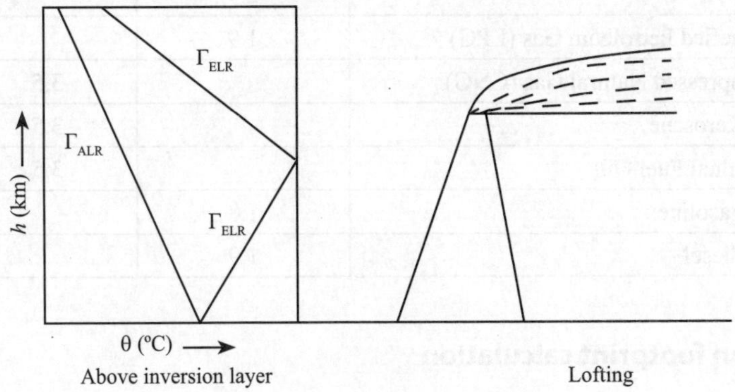

Fig. 3.16 Plume rise: Lofting.

Inference

It is suggested that smokestack be tall enough to emit plume above the inversion layer so that lofting plume is produced. Thus area near the smokestack is not polluted significantly. If the smokestack lies below the inversion layer, fumigation takes place and surrounding area will be heavily polluted.

3.7 CALCULATION OF CO_2 EMISSIONS

The Intergovernmental Panel on Climate Change (IPCC) provides a guideline to calculate CO_2 emission using oxidation factor. Generally the oxidation factor is taken as 0.99, since it is assumed that 99% carbon is oxidized to CO_2 and 1% remains unoxidized.

CO_2 emissions from a gallon of gasoline = 2,421 grams × 0.99 × (44/12) = 8,788 grams/gallon = 8.8 kg/gallon

CO_2 emissions from a gallon of diesel = 2,778 grams × 0.99 × (44/12) = 10,084 grams/gallon = 10.1 kg/gallon

CO_2 emission factors of different fuels are shown in the Table 3.4.

Table 3.4 Fuel emission conversion factor. 1 US gallon = 3.78541 litres.

Fuel type	*kg CO_2/litre*	*kg CO_2/kg*
Motor Gasoline	2.8	
Diesel Oil	2.9	
Gas Oil	2.9	
Liquefied Petroleum Gas (LPG)	1.9	
Compressed Natural Gas (CNG)		3.5
Jet Kerosene		3.5
Residual Fuel Oil		3.5
Biogasoline	1.8	
Biodiesel	1.9	

3.7.1 Carbon footprint calculation

Carbon footprint is defined as the amount of GHG emissions due to activity of an organization or person. Generally all GHG emissions are calculated as CO_2-equivalent using the concept of 100 year global warming potential or GWP 100.

Household energy

1. Electricity

$$CO_2 \text{ emissions in kg/year} = \frac{\text{average amount of electric bill per month}}{\text{price per KWH}} \times f \times \text{months/year}$$

f = Electricity emissions factor = 0.623 kg of CO_2/KWH

2. LPG

$$CO_2 \text{ emissions in kg/year} = \frac{\text{average amount of LPG bill per month}}{\text{price per kg of LPG}} \times f \times \text{months/year}$$

f = Natural gas emissions factor = 2.4 kg of CO_2/kg of LPG

3. Fuel oil

$$CO_2 \text{ emissions in kg/year} = \frac{\text{average amount of fuel bill per month}}{\text{price per gallon of fuel oil}} \times f \times \text{months/year}$$

f = Fuel oil emissions factor = 10.17 kg of CO_2/gallon of fuel oil

3.8 SOLVED PROBLEMS

Problem 1 Explain whether heating of Earth's surface by sunlight affect the stability of the atmosphere.

Solution Heating the ground warms the air and makes the lower atmosphere less stable.

Problem 2 Explain atmospheric stability and instability.

Solution If air is displaced vertically it will return to its position if the environment is ***stable*** and it will continue to move vertically in the direction of the displacement if the environment is ***unstable***.

Problem 3 What is meant by the term ***air parcel***?

Solution An air parcel is a small volume of air that is assumed to retain its identity as it moves through the atmosphere. It is assumed to be separated from surroundings by adiabatic wall.

Problem 4 What is common in the inflation of a tire and the descent of an air parcel?

Solution In both the cases compressed air is used and hence air is hot.

Problem 5 Why does an air parcel cool as it rises?

Solution As atmospheric pressure decreases with altitude, air parcels rising through the atmosphere expand and expansion requires work. Part of the thermal energy of the air is converted to expansion work, thereby decreasing the temperature of air parcel.

Problem 6 Which of the following is a constant? (i) dry adiabatic lapse rate, (ii) moist adiabatic lapse rate, (iii) environmental lapse rate.

Solution Only the dry adiabatic lapse rate is constant (10°C/km). The moist adiabatic lapse rate and environmental lapse rate vary with altitude.

Problem 7 Describe the motion of a vertically displaced air parcel in a stable atmosphere, an unstable atmosphere and a neutral atmosphere.

Solution When an air parcel is displaced vertically, it will return to its original position in a stable atmosphere but will move up continuously in an unstable atmosphere. In a neutral atmosphere when an air parcel is displaced vertically, it remains at its new position.

Problem 8 Why the moist adiabatic lapse rate is different from the dry adiabatic lapse rate?

Solution As moist air parcel rises, temperature decreases and condensation occurs with the release of latent heat, which increases the temperature of the air parcel. Thus actual temperature drop is always less than that of completely dry air parcel.

Problem 9 An inversion is an indicative of a stable layer in the atmosphere. Briefly explain why this is so.

Solution The environmental temperature increases with an increase in altitude within an inversion. As a parcel of air rises upwards, its temperature always decreases due to decompression. In an inversion, the environmental temperature will increase with altitude while the air parcel temperature decreases. Under this circumstance, if an air parcel rises, it will always be heavier than the surrounding environment and hence will return to its original position (stable).

Problem 10 Suppose the environmental lapse rate or ELR is approximately 4°C/km. Is it possible to develop strong convection at this ELR? Will it be possible to develop strong convection if ELR is 8°C/km?

Solution In the first case, the answer is No because saturated ALR is approximately 6°C/km and dry ELR is 10°C/km. So the rising air or air parcel will always be colder than its surrounding environment and return to its original position. So there is no question of development of strong convection.

If a saturated air or air parcel (ALR = 6ºC/km) is rising under ELR of 8ºC/km, condensation will occur and the parcel continues to rise since ELR > ALR. So a strong convective cloud will develop. However, if the lapse rate of air parcel is greater than 8ºC/km, ELR < ALR. So there is no possibility of development of strong convection.

Problem 11 What is convection?

Solution Convection describes buoyant vertical air currents within the atmosphere, often resulting in the formation of cumulus clouds. It occurs only when ELR > ALR.

Problem 12 Explain how atmospheric stability changes at a particular location.

Solution Atmospheric stability depends on ALR and ELR. ALR changes while heating or cooling the bottom of a layer. While heating warm air due to low specific gravity moves to the top layer and the reverse will occur while cooling.

Problem 13 What is the lifting condensation level?

Solution The lifting condensation level is that altitude at which condensation first occurs in a lifted air parcel.

Problem 14 What is the level of free convection?

Solution It refers to that altitude, at which ELR is just greater than ALR and the air parcel starts rising on its own.

Problem 15 If the surface winds are convergent or divergent can air parcel move upwards in that location?

Solution Under convergent condition only air parcel can move upwards. When the surface winds are convergent, resultant air starts rising upwards due to excess air pressure. When the surface winds are divergent, pressure is low at the bottom. So air moves down from above.

Problem 16 What is the Lifted Index? How is the value of the Lifted Index related to the severity of storms?

Solution The Lifted Index is a measure of instability of the atmosphere. It is calculated by subtracting the temperature of a parcel lifted to 500 mbar (from the surface) from the temperature of the environment at 500 mbar. The lower the value of Lifted Index, the more potential for severe thunderstorms.

Problem 17 A balloon is lifted at an altitude of 1 km from the surface of the Earth. Inside temperature of the balloon is 25°C and outside temperature of the balloon is 2°C. Assume that the balloon is filled up with saturated air. The environmental lapse rate is 5°C/km. Estimate the highest altitude at which the balloon can rise.

Solution The balloon will rise till ELR > ALR. When ELR = ALR, the balloon stops rising and starts coming down. Suppose Z is the highest altitude that the balloon can rise. As the balloon is filled up with saturated air, ALR is 10°C/km.

Temperature inside the balloon at a height $Z = T_1 = 25°\text{C} - 10°\text{C/km} \times Z$
Temperature outside the balloon at a height $Z = T_2 = 2°\text{C} - 5°\text{C/km} \times Z$
So, $T_1 = T_2$ at the height Z. Hence, $Z = 4.6$ km.
As the balloon is lifted at an altitude of 1 km from the surface of the Earth, the highest altitude at which the balloon can rise is (4.6 + 1) = 5.6 km.

Problem 18 Consider air is flowing onshore at the surface from the Pacific Ocean with a temperature of 10°C and a dewpoint of 5°C. The onshore airflow is forced to ascend the mountains, where its trajectory reaches a maximum elevation of 3 km. The air then descends to the plains, where the elevation is 1.5 km. Assume that the air's dewpoint decreases at 2°C km of ascent prior to saturation, increases at the same rate during descent and decreases at the moist adiabatic lapse rate of 6°C/km during saturated ascent.

(a) Calculate the relative humidity of the air as it comes onshore. Given saturated vapour pressure at 10°C is 12.3 mbar and at 5°C is 8.7 mbar.
(b) Determine the temperature, dewpoint and relative humidity of the air at the top of the mountains (3 km).
(c) Determine the temperature, dewpoint and relative humidity of the air when it reaches the plains (1.5 km).
(d) Suppose the air descended farther to sea level. What would be its temperature and relative humidity? Explain why the values are different from when the air first came onshore.

Solution (a) Relative Humidity = (8.7/12.3) × 100 ≈ 70%

(b) First find the altitude where $T = T_d$.

$T = 10°C/km \times Z$, $T_d = 5°C - 2°C/km \times Z$

Hence, Z = 0.625 km. Using this value of Z, $T = 3.75°C$

Thus temperature at the mountain top $T = 3.75°C - 6°C/km \times (3 - 0.625) = -10.5°C$.

Air is saturated at 3 km, so $T = T_d$ and RH = 100%.

(c) $T = -10.5°C + 10°C/km \times 1.5 = 4.5°C$

$T_d = -10.5°C + 2°C/km \times 1.5 = -7.5°C$

RH = 41%

(d) $T = -10.5°C + 10°C/km \times 3 = 19.5°C$.

$T_d = -10.5°C + 2°C/km \times 3 = -4.5°C$

RH = 20%. It is well below 70% because moisture was removed as precipitation occurs during ascent over mountains.

Problem 19. Combustion of methane provides about 8.2 × 10¹⁶ kJ/year. Energy content of methane is 39 × 10³ kJ/m³ at STP. Calculate the emission rate of carbon in the atmosphere due to combustion of methane.

Solution

$$\text{moles of } CH_4 = \frac{8.2\times10^{16}\,\text{kJ/year}}{39\times10^{3}\,\text{kJ/m}^3} \times \frac{1}{22.4\times10^{-3}\,\text{m}^3/\text{mole}}$$

$$= 9.4 \times 10^{13}\ \text{mole/year}$$

1 mole CH_4 gives rise 1 mole CO_2 and hence 12 gm C.

9.4 × 10^{13} mole CH_4/year gives rise 9.4 × 10^{13} mole CO_2/year, which is equivalent to ≡ 9.4 × 10^{13} × 12 gm C/year = 1.1 × 10^9 tonnes/year = 1.1 Gt/year.

Thus the emission rate of carbon in the atmosphere is 1.1 Gt/year.

4

Greenhouse Effect

4.1 SURFACE TEMPERATURE OF EARTH

4.1.1 Black body model

4.1.1.1 Assumptions

(i) The Earth is considered as a black body.

(ii) The Earth surface is considered isothermal with an average equivalent temperature T_e °K.

(iii) The Earth is considered as a sphere of radius R.

(iv) The Earth absorbs all the radiations entering into its atmosphere.

The projected surface area of the Earth is πR^2. The solar flux entering into Earth's atmosphere is S Watt/m². It is also called insolation. Thus net energy, absorbed by the Earth's surface is $S \times \pi R^2$ Watt.

Amount of energy, radiated by the Earth's surface, may be calculated by using Stephen-Boltzmann law, according to which

Amount of energy radiated = $\sigma \times 4\pi R^2 \times T_e^4$ Watt, where, σ is known as Stephen-Boltzmann constant and its value is 5.67×10^{-8} W/m². $(K)^4$.

As the Earth is considered as black body, we can write that

Amount of energy absorbed = Amount of energy radiated.

$$S \times \pi R^2 = \sigma \times 4\pi R^2 \times T_e^4,$$

$$T_e = \left(\frac{S}{4\sigma}\right)^{\frac{1}{4}} K = \left(\frac{1372}{4 \times 5.67 \times 10^{-8}}\right)^{\frac{1}{4}} K = 279 \text{ K}$$

since the value of S in case of Earth is 1372 Watt/m². The value of S is different for different planets since it depends on the distance (d) between the Sun and the planet. The following relation is assumed to hold good

$$S \propto \frac{1}{d^2}$$

The observed average surface temperature of the Earth is 288 K, which is close to the theoretical value but still the theoretical calculation is inaccurate since the model does not consider the Albedo.

4.1.1.2 Albedo

In actual practice the Earth cannot absorb all the radiations entering into its atmosphere. A part of the solar flux is reflected back to the space by the Earth's atmosphere and the Earth's surface as well. This reflected fraction is called Albedo and its value varies from planet to planet. In case of the Earth the value of albedo is approximately 0.3. Albedo is represented by the symbol α. It is shown in the Fig. 4.1.

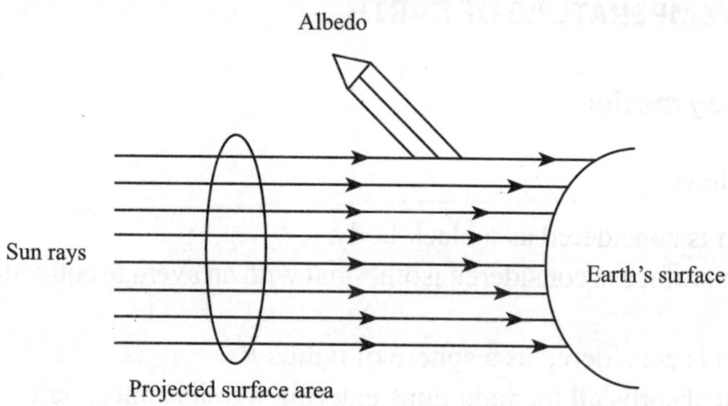

Fig. 4.1 Solar flux entering into Earth's atmosphere.

Considering the albedo the above equation can be rearranged as follows:

$$S(1-\alpha)\times \pi R^2 = \sigma \times 4\pi R^2 \times T_e^4$$

$$T_e = \left[\frac{S(1-\alpha)}{4\sigma}\right]^{\frac{1}{4}} K = \left(\frac{1372\times 0.7}{45.67\times 10^{-8}}\right)^{\frac{1}{4}} K = 255 \text{ K}$$

Thus the difference between average surface temperature of Earth and theoretical surface temperature of Earth is (288 – 255) K = 33 K = 33°C. This difference in temperature is assumed to arise due to Greenhouse Effect.

4.1.2 Wien's displacement law

If a black body is heated up by monochromatic beam of light, it initially absorbs all the radiations, striking on its surface but later on it emits the absorbed radiations in a wide range of spectrum as shown in the Fig. 4.2.

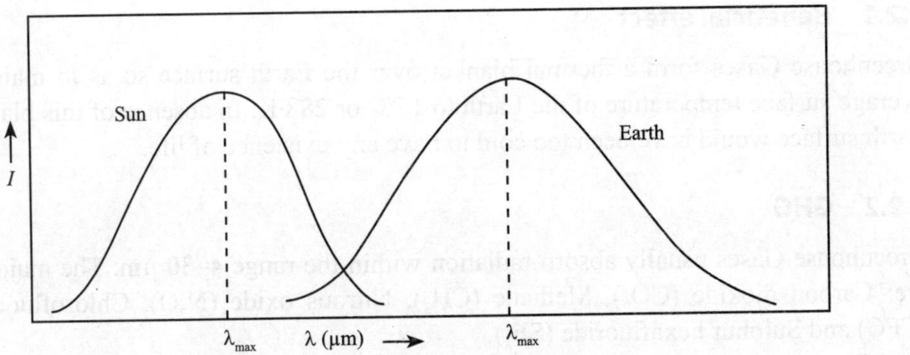

Fig. 4.2 Spectrum of emitted radiations from the Earth and the Sun.

The spectrum shows a maximum intensity and the corresponding wavelength is called λ_{max}. If λ_{max} is expressed in μm and surface temperature of the black body is expressed in °K, the following equation is true for all black bodies

$$\lambda_{max} \times T = 2898$$

Or,
$$\lambda_{max} = \frac{2898}{T}$$

The above relation is popularly known as Wien's displacement law. By using this law surface temperature of any black body can be calculated by measuring λ_{max}.

In the case of the Sun the λ_{max} value is 0.5 μm and hence the surface temperature of the Sun is given by

$$T = \frac{2898}{\lambda_{max}} = \frac{2898}{0.5} = 5796 \text{ K}$$

In the case of Earth the λ_{max} value is 10.1 μm and hence the surface temperature of the Earth is given by

$$T = \frac{2898}{\lambda_{max}} = \frac{2898}{10.1} = 286.93 \text{ K}$$

4.2 GREENHOUSE EFFECT

As the Earth is considered as a black body, it emits all the absorbed radiation. Part of this emitted radiation easily passes through the atmosphere to space and the remaining part of the emitted radiation is entrapped within the atmosphere. Thus atmosphere acts like a glass window, known as atmospheric radiative window. Eventually the trapped radiation returns to the Earth surface, thereby increasing the surface temperature of the Earth from 255 K to 288 K. This is known as greenhouse effect. The gases, responsible for trapping radiation, are called Greenhouse Gases or GHGs. Wavelength of trapped radiation usually lies within 4–30 μm.

4.2.1 Beneficial effect

Greenhouse Gases form a thermal blanket over the Earth surface so as to maintain the average surface temperature of the Earth to 15ºC or 288 K. In absence of this blanket, the Earth surface would have been too cold to have any existence of life.

4.2.2 GHG

Greenhouse Gases usually absorb radiation within the range 4–30 μm. The major GHGs are: Carbon dioxide (CO_2), Methane (CH_4), Nitrous oxide (N_2O), Chlorofluorocarbon (CFC) and Sulphur hexafluoride (SF_6).

There are some minor GHGs, e.g., ozone, sulphur dioxide, water vapour etc.

Table 4.1 Characteristic features of major GHGs.

GHG	*Average concentration in atmosphere*	*Atmospheric life time (years)*	*GWP calculated over 100 years*	*Contribution in global warming*
CO_2	390 *ppm*	Not known	1	50%
CH_4	1.8 *ppm*	11	25	19%
N_2O	0.4 *ppm*	150	300	4%
CFC	5×10^{-4} *ppm*	400	15000	17%
SF_6	Trace amount	3200	23000	5%

As Global Warming Potential (GWP) is calculated with respect to CO_2, GWP of CO_2 is 1. Water vapour has a significant influence in absorbing infrared radiation. However, its GWP is not calculated since its concentration in the atmosphere varies widely and mainly depends on air temperature.

4.3 ADVERSE EFFECTS OF GHG

(a) The concentration of carbon dioxide in the Earth's atmosphere increases continuously, thereby increasing average surface temperature of the Earth. As a result of this, evaporation rate of water from ocean and other water reservoir increases and hence moisture content of the atmosphere increases. As water vapour itself is a GHG, this increased moisture content further increases the surface temperature of the Earth.

(b) (i) One important parameter for food-grain production is temperature. Due to global warming atmospheric temperature rises. Food-grain production is affected with increase in temperature even by 1ºC.

(ii) Due to global warming atmospheric temperature increases and hence surface temperature of water increases. So dissolved oxygen content decreases, threatening the aquatic ecosystem.

(iii) As GHG concentration in the atmosphere increases, surface temperature of ocean water increases significantly, as a result of which, transfer of nutrients from the bottom layer of ocean to the surface is prohibited.

4.4 GLOBAL WARMING

4.4.1 Global warming potential (GWP)

For a given GHG, it refers to the amount of heat trapped by 1 kg of that GHG with respect to that by 1 kg of carbon dioxide over a certain period of time. Time period may be 20, 100 or 500 years. Thus GWP is a ratio and has no unit. GWP of a greenhouse gas depends on the following factors

(a) Capacity of absorption of infrared radiation.

(b) The wavelength window of absorbed infrared radiation.

(c) The atmospheric lifetime of the greenhouse gas.

Thus, a high GWP correlates with a large infrared absorption and a long atmospheric lifetime.

4.4.1.1 Significances

1. Global warming impact of any GHG can be realized from the knowledge of its GWP.
2. It also indicates that GHG, which has longer atmospheric life time, has higher global warming impact.

4.4.1.2 Calculation of GWP

GWP of a GHG is calculated on the basis of radiative forcing capacity or radiative efficiency (RF) of that GHG. RF is the amount of energy per unit area, per unit time, absorbed by the greenhouse gas that would otherwise be lost to space. Radiative forcing is a measure of the influence of a factor that is altering the balance of incoming and outgoing energy in the Earth-atmosphere system. It is an index of importance of the factor that has potential to change the climate. In this report radiative forcing values are for changes relative to preindustrial conditions defined at 1750 and are expressed in Watt per square meter (W/m^2).

Radiative forcing can be used to estimate a subsequent change in equilibrium surface temperature (ΔT_s) arising from that radiative forcing via the equation

$$\Delta T_s = \lambda.\Delta F$$

where λ is the climate sensitivity, usually with units in $^{o}K/(W/m^2)$ and ΔF is the radiative forcing. A typical value of λ is $0.8^{o}K/(W/m^2)$, which gives a warming of 3 K for doubling of CO_2 concentration.

4.4.1.3 Equivalent carbon dioxide

Radiative forcing of different GHGs can be expressed as RF-CO_2-eq. It is expressed as parts per million by volume or *ppmv*. RF-CO_2-eq. is expressed as

$$RF = \alpha \ln \frac{C}{C_0}$$

α is a constant and its value is 5.35. C_0 is the reference concentration of CO_2 and C is the present concentration of GHG, also expressed as CO_2-eq.

For example reference concentration of $CO_2 = C_0 = 290$ *ppmv* in 1750. The RF values of CO_2, CH_4, NO_2 are 1.46, 0.48 and 0.15 W/m^2 respectively. The RF value of other minor gases is 0.01 W/m^2. So total RF = 1.46 + 0.48 + 0.15 + 0.01 = 2.1 W/m^2. Thus CO_2-eq. is calculated using the reference concentration of CO_2 as 290 *ppmv* as below

$$C = C_0 e^{RF/\alpha} = 290 \times e^{2.1/5.35} = 429.41 \; ppmv$$

According to IPCC (Intergovernmental Panel on Climate Change) the GWP is defined as the ratio of the time-integrated radiative forcing from the instantaneous release of 1 kg of a trace substance relative to that of 1 kg of a reference gas

$$GWP\,(x) = \frac{\int_0^T a_x \cdot [x_t]\,dt}{\int_0^T a_r \cdot [r_t]\,dt}$$

where T is the time horizon over which the calculation is considered.

a_x = The radiative efficiency due to a unit increase in atmospheric abundance of the GHG in Wm^{-2} kg^{-1}.

a_r = The radiative efficiency due to a unit increase in atmospheric abundance of CO_2 in Wm^{-2} kg^{-1}.

$[x_t]$ = The time-dependent decay in abundance of the GHG immediately after an instantaneous release of it at time $t = 0$.

$[r_t]$ = It is the time-dependent decay in abundance of CO_2 immediately after an instantaneous release of it at time $t = 0$.

However, a_x and a_r may not be constant over time period considered.

4.4.2 Steps to be taken to control global warming

(a) Consumption of fossil fuel, e.g., coal, petroleum, should be reduced.

(b) Non-conventional energy resources should be commercialized. Use of biofuel is recommended.

(c) Culture of plantation of trees should be spread globally to increase the CO_2 consumption.

(d) Emission of GHG should be restricted.

(e) All nations should abide by different international protocols regarding global warming.

4.4.3 Kyoto protocol

The 1st major international step to prevent global warming was taken at Rio de Jenerio, Brazil in 1992, where 192 nations adopted the United Nations Framework Convention on Climate Change (UNFCCC). It was recognized that reduction in CO_2 and other GHG emissions is necessary for preventing catastrophic climate change. The major achievement of the UNFCCC process was the Kyoto Protocol, adopted on December 11, 1997. Under this protocol, 187 nations excepting the United States, agreed to reduce their CO_2 emission to 5.2% of their 1992 levels at the end of 2012.

- Industrialized nations reduce CO_2 emission to 5 percent from their 1990 levels by 2008-2012 compliance period.
- United States withdrew in 2001.
- China and India are not required to comply because they are developing nations.
- By 2002 Kyoto only covered about 30 percent of global CO_2 emission.

4.5 EL-NINO AND LA-NINA

El-Nino is an inter annual fluctuation in the ocean atmospheric system. It disrupts normal weather around the pacific ocean. As a result of this heavy rain and flood are observed in countries like Peru and Australia faces droughts. It is usually observed every 10 years. Usually wind is blowing westwards and nutrient rich bottom water moves on to the surface. The water region near the surface is warmer than that at the bottom. Due to global warming sufficient temperature difference is developed between surface water and bottom water. This local phenomena of warming of ocean water near South America is called El-Nino.

This warm water near South America is then streaming to the western pacific region, bringing about intense storm that hits Australia and part of Western Asia, causing droughts in North America and floods in Bangladesh, Sudan. This phenomena is called La-Nina.

4.6 RADIATION BALANCE

Solar radiation entering into the Earth's surface is $1372 \times \pi R^2$ Watt and amount of solar flux entering into Earth's surface is $1372 \times \pi R^2/4\pi R^2 = 343$ Watt/m^2. However, part of this solar flux, i.e., 0.3×343 units ≈ 103 units is reflected back to the space. Out of this 103 units, 89 units are reflected by Earth's atmosphere and 14 units are reflected by Earth's surface. Remaining 240 units are absorbed, out of which, 86 units are absorbed within the atmosphere and 154 units are absorbed on the Earth's surface.

However, some radiations are emitted by Earth's surface and absorbed within the atmosphere.

(i) 80 units are emitted from the Earth's surface as latent heat of water.
(ii) 16 units are emitted from the Earth's surface as sensible heat other than latent heat.
(iii) 372 units are radiated from the Earth's surface to the atmosphere.
(iv) 18 units of radiation passes to the atmosphere to space.

(v) 332 units of radiation are returned to the Earth's surface from the atmosphere.
(vi) 222 units of radiation are emitted from the atmosphere to space.

The total radiation balance is shown in the Fig. 4.3.

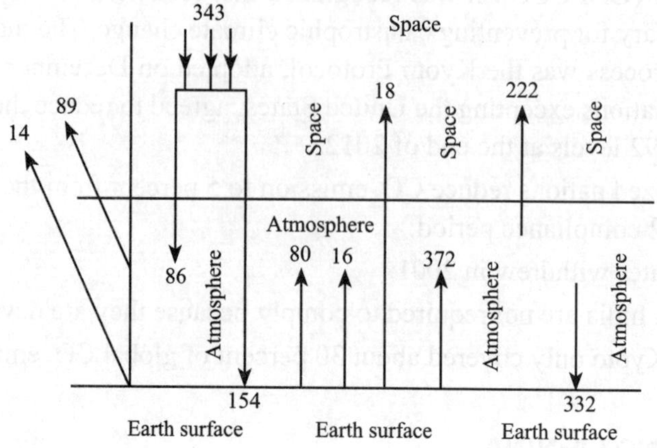

Fig. 4.3 Radiation balance by Earth's surface, atmosphere and space.

Radiation balance on Earth's surface

Total incoming radiation	(154 + 332) unit = 486 units
Total outgoing radiation	(80 + 16 + 18 + 372) = 486 units

Radiation balance on Earth's atmosphere

Total incoming radiation	(80 + 16 + 86 + 372) = 554 units
Total outgoing radiation	(332 + 222) = 554 units

Radiation balance on space

Total incoming radiation	(18 + 222) = 240 units
Total outgoing radiation	(86 + 154) = 240 units

4.7 NUMERICAL PROBLEMS

Problem 1 Consider the Earth is flat and an isothermal body. Albedo for the Earth is 0.2. Estimate the temperature of the flat Earth.
Given: $S = 1372\ W/m^2$, $\sigma = 5.67 \times 10^{-8}\ W/m^2\ (K)^4$.

Solution As the Earth is flat the surface area of the Earth is $2 \times \pi R^2$. The factor 2 is introduced because flat Earth has two surfaces.
As the Earth is considered as black body, we can write that
Amount of energy absorbed = $S(1 - \alpha) \times \pi R^2 = 1372 \times 0.8 \times \pi R^2$ Watt

Amount of energy radiated = $\sigma \times 2\pi R^2 \times T_e^4 = 5.67 \times 10^{-8} \times 2\pi R^2 \times T_e^4$ Watt

Hence,

$1372\,(1 - 0.2) \times \pi R^2 = 5.67 \times 10^{-8} \times 2\pi R^2 \times T_e^4$.

$$T_e = \left(\frac{1372 \times 0.8}{2 \times 5.67 \times 10^{-8}}\right)^{\frac{1}{4}} K = 313.6 \text{ K}$$

Problem 2 Solar flux or insolation (S), striking on a planet, is given by:

$S \propto \frac{1}{d^2} = \frac{k}{d^2}$, where, d is the distance between the Sun and the planet.

(i) Find k given d for the Earth is 150×10^9 m and $S = 1372$ W/m².

(ii) Estimate d for the Venus. Given: $S = 2613$ W/m²

(iii) Estimate S for the Mercury. Given: $d = 58 \times 10^9$ m

Solution (i) $1372 = \frac{k}{\left(150 \times 10^9\right)^2}$. Hence, $k = 3.087 \times 10^{25}$ Watt.

(ii) $d^2 = \frac{k}{S} = \frac{3.087 \times 10^{25}}{2613} \text{ m}^2 = \frac{30870000 \times 10^{18}}{2613} \text{ m}^2 = 11814 \times 10^{18} \text{ m}^2$

$d = 109 \times 10^9$ m.

(iii) $S = \frac{k}{d^2} = \frac{3.087 \times 10^{25}}{\left(58 \times 10^9\right)^2} = \frac{30870000}{(58)^2} = 9176$ Watt/m²

Problem 3 Estimate equivalent temperature and surface temperature of the Mars by using the following data:

$S = 589$ W/m², $\alpha = 0.15$, $\sigma = 5.67 \times 10^{-8}$ W/m² (K)⁴, $\lambda_{max} = 13.173$ µm

Solution We know that

$$T_e = \left[\frac{S(1-\alpha)}{4\sigma}\right]^{\frac{1}{4}} K = \left[\frac{589(1-0.15)}{4 \times 5.67 \times 10^{-8}}\right]^{\frac{1}{4}} K = 217 \text{ K}$$

λ_{max} (µm) × T_s (°K) = 2898.

Or, $T_s = \frac{2898}{13.173} K = 220$ K

Problem 4 The solar flux at the outer edge of the atmosphere of the Earth varies by ±3.3%.

Calculate the change in equivalent temperature.

Given: $\alpha = 0.3$, $S = 1372$ W/m², $\sigma = 5.67 \times 10^{-8}$ W/m² (K)⁴.

Solution Equivalent temperature of Earth's surface is given by

$$T_e = \left[\frac{S(1-\alpha)}{4\sigma}\right]^{\frac{1}{4}} K = \left(\frac{1372 \times 0.7}{4 \times 5.67 \times 10^{-8}}\right)^{\frac{1}{4}} K = 255 \text{ K}$$

At one edge of the atmosphere of the Earth, solar flux is $S(1 + 0.033)$. So the corresponding equivalent temperature is given by

$$T_e^1 = \left[\frac{S(1-\alpha)}{4\sigma}\right]^{\frac{1}{4}} K = \left(\frac{1372 \times 1.033 \times 0.7}{4 \times 5.67 \times 10^{-8}}\right)^{\frac{1}{4}} K = 257 \text{ K}$$

At the other edge of the atmosphere of the Earth, solar flux is $S(1 - 0.033)$. So the corresponding equivalent temperature is given by

$$T_e^2 = \left[\frac{S(1-\alpha)}{4\sigma}\right]^{\frac{1}{4}} K = \left(\frac{1372 \times 0.967 \times 0.7}{4 \times 5.67 \times 10^{-8}}\right)^{\frac{1}{4}} K = 253 \text{ K}$$

$$\Delta T_e = T_e^1 - T_e^2 = (257 - 253)\,K = 4 \text{ K}$$

5

Noise Pollution

5.1 INTRODUCTION

5.1.1 Definition of noise

Any sound, which causes irritation to the human ear, is called noise. Two important parameters of noise are (i) sound intensity and (ii) sound pressure.

5.1.2 Measurement of noise

5.1.2.1 Sound intensity level

Sound intensity level can be measured by standardizing sound intensity or sound pressure. Sound pressure is measured in micro Pascals. The range of audible sound is from 20–200,000,000 micro Pascals. The dBSPL scale is a logarithmic scale using sound pressure, sound pressure increases ten times with increase in sound intensity by 20 decibel. The Standard reference value of sound intensity detectable at 1 KHz is 20 micro Pascals and corresponding sound intensity is $I_0 = 10^{-12}$ W/m². This is called 0 dBSPL. Noise level is usually expressed as deciBel (dB), which is given by:

$$\text{dB} - \text{scale} = 10 \log \frac{I}{I_0}$$

where, I = Measured intensity and I_0 = Reference Intensity.

Acoustic design and sound measurement depends on human ear and hence sound meters are normally fitted with filters adapting the measured sound response to the human sense of sound. Common filters are (i) dB(A), (ii) dB(B), dB(C).

The decibel A filter is widely used. The value of dB(A) is approximately zero at sound intensity of 40 dB at 1 kHz equal-loudness curve for the human ear. Using the dB(A)-filter, the sound level meter is less sensitive to very high and very low frequencies. Sound intensity, measured with this scale, is expressed as dB(A).

The decibel C filter is almost linear over several octaves and hence suitable for sound intensity measurements at very high sound pressure levels. The decibel B filter lies between A and C. However, B and C filters are rarely used. It is illustrated in the Fig. 5.1.

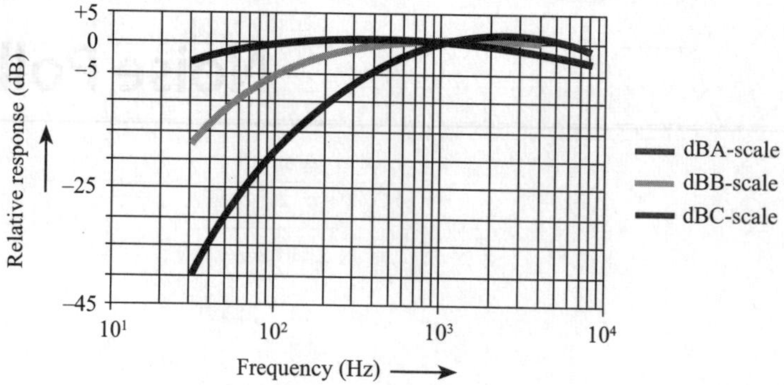

Fig. 5.1 Different loudness scales of noise.

Relative Response (dB)	*Frequency (Hz)*								
	31.5	63	125	250	500	1000	2000	4000	8000
dB(A)	–39.4	–26.2	–16.1	–8.6	–3.2	0	1.2	1	–1.1
dB(B)	–17	–9	–4	–1	0	0	0	–1	–3
dB(C)	–3	–0.8	–0.2	0	0	0	–0.2	–0.8	–3

Octave	1	2	3	4	5	6	7	8
Measured sound pressure level (dB)	54	60	64	53	48	43	39	32
dB(A) filter	26	16	9	4	0	–1	–1	1
Resulting sound pressure level (dB)	28	44	55	49	48	44	40	31

Sound pressure level (dBSPL)

Standard reference value of pressure intensity = p_0 = 2 × 10^{-5} N/m^2. As intensity is proportional to the square of the sound pressure, the corresponding noise level may be expressed as sound pressure level or dBSPL, which is given by

$$\text{dBSPL} = 20 \log \frac{p}{p_0}$$

where, p = measured pressure intensity.

dBSPL scale is used for hearing aid performance. dBSPLvalues at different sound pressures with examples are listed in Table 5.1.

Table 5.1 dBSPL values at different sound pressures.

Sound pressure (μPa)	*dBSPL at 1 kHz*	*Example*
20	0	Just audible
200	20	Rustle of leaf
2,000	40	Quiet speech
20,000	60	Conversational speech
2,00,000	80	A shout
2,000,000	100	Pneumatic drill
20,000,000	120	Aircraft taking off
200,000,000	140	Rocket launching

Zero dB When the measured sound intensity is same as the reference sound intensity, i.e., 2×10^{-5} N/m^2, the resultant sound intensity is zero. It is called zero dB.

The dB-scale or dBSPL is used to measure general noise level but the scale is not sufficient to measure peak noise level, duration of noise level and quality of noise. The ISO has introduced several noise-meters, based on dB-scale, to avoid this difficulty, i.e., L_{10}, L_{eq}, Lepn.

5.1.2.2 Sound power level

The relationship between sound power level (L_w) and sound intensity level (or sound pressure level) is given by the following relationship

$$L_w = 10 \log \frac{I}{I_0} + 10 \log \frac{\alpha}{\alpha_0},$$

$$L_w = 10 \log \frac{W}{W_0},$$

$$L_p = 10 \log \frac{p^2}{p_0^2}$$

where, α = Surface area of sphere in m^2 and α_0 = Reference area = 1 m^2.

p = Root mean square (r.m.s.) sound pressure in N/m^2.

p_0^2 = Reference root mean square (r.m.s.) sound pressure in N/m^2.

Thus, $L_w = L_I + 10 \log \alpha$

5.1.2.3 L_{10}(18 hour) Index

It is the arithmetic average of hourly values of noise level between 6 A.M. to 12 Midnight on any normal working day. It is particularly useful to measure road traffic noise.

5.1.2.4 Effective Perceived Noise Level (*Lepn*)

This index of noise level is usually used to measure peak value of noise level during taking off or landing of aeroplanes at different airports. Lepn = dBA – scale + 13. It is introduced by International Civil Aviation Organization (ICAO).

5.1.2.5 Equivalent Noise Level (L_{eq})

This index is particularly useful to measure a mean noise level over a specified time period. For example, L_{eq} = 90 dBA means an average noise level of 90 dBA over a specified time period. It also indicates a steady noise level at 93 dBA for half of that specified time period and zero noise level for the other half of that time period. It is recommended by International Organization for Standardization (ISO).

5.1.2.6 Phon and sone

Decibel scale is used to express instantaneous sound intensity. However, sound intensity level depends on frequency and the decibel unit does not consider the frequency factor.

Phon It is also used to express sound intensity level but it considers the frequency factor. Human ear is less sensitive to a vibration above or below 1000 Hz. Thus above or below 1000 Hz, loudness decreases. Considering the above fact another noise unit phon has been introduced. It is the sound intensity level, in dB unit, measured at 1000 Hz. Suppose at a given instant the sound intensity is measured 50 dB at 1000 Hz. Then the corresponding sound intensity level is 50 phons. If the same sound intensity is measured at different frequency, the corresponding phon unit is different. It is also expressed in log scale. So phon unit of noise does not have additive property, that means, 60 phons + 60 phons ≠ 120 phons.

Sone It is also a measure of loudness. It is the most widely used to express noise intensity level since it follows additive rule, that means 1 sone + 1 sone = 2 sones. A sound level of 40 dB at 1000 Hz is called 1 sone. The relation between phons and sones is $S = 2^{(P-40/10}$, where, P and S represent phons and sones respectively.

$$\log S = \left(\frac{P-40}{10}\right)\log 2$$
$$= 0.030103\, P - 1.20412$$

The plot of log S versus P is shown in the Fig. 5.2.

When, P = 40, S = 1 and P = 50, S = 2. Thus two noises of intensity 40 phons each, produce 2 sones or 50 phons, not 80 phons. Thus sone but not phon has additive characteristics.

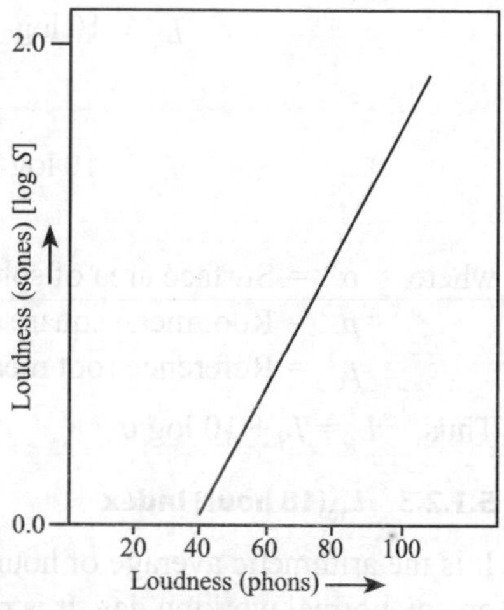

Fig. 5.2 Relation between phons and sones.

5.2 NOISE DOSE

The exposure limit of a specific noise level that a person can tolerate without any damaging effect is called noise dose limit. It depends on loudness of sound. Less is the noise level, higher is the noise dose limit. The committees (OSHA and CHABA) have jointly set allowable exposure times for different noise levels. The noise dose limit is listed in Table 5.2.

OSHA Occupational Safety and Health Act
CHABA Committee on Hearing, Bioacoustics and Biomechanics Act

Table 5.2 Permissible noise dose time limit for different sound level.

Sound level (dBA)	*Time (hrs.)*	*Sound level (dBA)*	*Time (hrs.)*
85	16.00	101	1.73
86	13.93	102	1.52
87	12.13	103	1.32
88	10.57	104	1.15
89	9.18	105	1.00
90	8.00	106	0.87
91	6.97	107	0.77
92	6.07	108	0.67
93	5.28	109	0.57
94	4.60	110	0.50
95	4.00	111	0.43
96	3.48	112	0.38
97	3.03	113	0.33
98	2.83	114	0.28
99	2.25	115	0.25
100	2.00		

5.3 CLASSIFICATION OF NOISE

Noise can be classified into three broad categories

(i) Transport noise, (ii) Occupational noise, (iii) Neighbourhood noise.

5.3.1 Transport noise

It may be further divided into three categories

(i) Road traffic noise, (ii) Rail traffic noise, (iii) Aircraft noise.

5.3.1.1 Road traffic noise

Noise, produced from automobiles and other vehicles, is called road traffic noise. So noise level increases with increase in traffic density. Permissible sound level in different areas is listed below

Table 5.3 Permissible noise level in different areas.

Areas	*Day time*	*Night time*
Industrial	75 dB	65 dB
Commercial	65 dB	55 dB
Residential	50 dB	45 dB
100 m around hospitals, educational institutions, courts etc.	45 dB	40 dB

5.3.1.2 Rail traffic noise

The frequency of rail traffic noise is lower than that of road traffic noise and hence apparently it is not harmful to human being. However, buildings, located nearby railway tracks, are affected by that noise level. Introduction of diesel and all-electric locomotives has greatly reduced rail traffic noise.

5.3.1.3 Aircraft noise

Unlike road traffic noise, aircraft noise is not continuous. This type of noise arises only when aeroplanes are landing or taking off. Permissible noise level range is 100 Lepn to 120 Lepn in day time and 90 Lepn to 100 Lepn at night.

5.3.2 Occupational noise

It arises mainly due to (i) industrial machines and (ii) domestic machines.

Examples of industrial machines are Milling machine, Lathe machine, Newspaper Press, Textile Loom etc.

Examples of domestic machines are Vacuum Cleaner, Washing machine, Mixer & Grinder etc.

5.3.3 Neighbourhood noise

The noise created during individual or public entertainment is called neighbourhood noise.

Television-set, Radio-set, Loudspeaker, Fire Crackers etc. are some of the common sources, which cause huge disturbances to general public.

5.4 ADVERSE EFFECTS OF NOISE POLLUTION

(a) Temporary or permanent damage to hearing mechanism.
(b) Causes annoyance and complaints.

(c) Very low level noise impairs verbal communications. For conversation within a metre, 65 dBA is enough, 125 dBA gives the sensation of pain and 150 dBA might kill a human being. (Physical effect).

(d) Persisting noise level above 90 dBA for more than 10 min reduces the sensitivity of ear. High frequency noise can affect the semi-circular canals of inner ear and affected person may suffer from nausea and dizziness. It also disturbs verbal communication.

(e) Audible frequency noise can generate resonance in the skull and hence affect the brain and nervous system.

(f) Moderate frequency noise may lead to pain, numbness and cyanosis of fingers.

(g) Very high frequency noise causes damage to bones and joints of human body.

(h) Very low frequency noise may produce resonance in the body organs, giving rise to reduced heart beat, reduced vibrations in blood pressure and breathing difficulties.

(i) Low frequency noise (less than 40 dBA) causes disturbances to brain cells and changes the normal sleep pattern. (Physiological effect).

(j) Sound pollution is the prime cause of physical and mental fatigue and loss of concentration, depression, disturbances in verbal communication, change of sleep pattern. (Psychological effect)

5.5 CONTROL OF NOISE POLLUTION

In the simplest form noise can be controlled by maintaining the following guidelines

(a) Setting industries far away from residential areas.

(b) Proper and well-designed engines should be used to reduce the noise level.

(c) Restricting the use of loudspeakers and amplifiers.

(d) Planting trees along both sides of roads.

(e) Acoustic materials should be used to absorb sound.

Noise is always generated from a source and then it is communicated to the receiver end. So, noise can be controlled at source, along communicating path and also at the receiver end.

(i) At source

(a) *Proper design* shock absorption technique. Using silencer in automobiles. Efficient fluid flow technique. Reducing sound radiating areas.

(b) *Proper operation* acoustic guards, proper cutting speed, feeding rate, additional sound absorbing device.

(c) *Equipment maintenance* Proper lubrication, purchasing and regular replacing of components, proper cleaning, proper installation.

(ii) Along pathways

(a) Disseminating vibration energy, transmission of noise through corridors, walls etc.

(b) Formation of multiple layers between source and receiver so as to reflect noise in different directions.

(c) Proper insulation and using sound absorbing materials surrounding machines.

(iii) At receiver

(a) Using ear plugs.

(b) Using sound dampers.

(c) Using filters which absorb high and low frequency noise.

(d) Using acoustic materials in walls and ceiling.

5.6 THE APPLICATIONS OF ULTRASONIC SOUND WAVES

The ultrasonic sound waves (frequency >20 kHz) is used to detect different submerged rocks, sunken ships, submarines etc. in the ocean. It is also used to measure the depth of the ocean. It is widely used in medical purposes, for example in ultrasonographic equipment to get images of different parts of human body.

5.7 NOISE THRESHOLD LIMIT VALUE (NTLV)

A worker can tolerate a given noise level up to a certain period without any adverse effect on his health. This duration of noise level is known as limit value of that noise level. The following chart illustrates the limit value of different noise levels. This is also known as noise dose limit. If more than one noise level is active in a given system, combined effect should be considered. When the sum of ratios of noise level durations to the corresponding limit values is equal to 1, the combined effect reaches the noise threshold limit value. When the sum is greater than 1, the combined effect exceeds the noise threshold limit value and the situation is unhealthy.

$$\frac{C_1}{T_1}+\frac{C_2}{T_2}+\frac{C_3}{T_3}+\cdots\cdots\cdots\cdots=1 \text{ (NTLV)}$$

Different noise levels with maximum tolerance limits or noise dose limits are listed in Table 5.4.

Table 5.4 Different noise levels with maximum tolerance limit.

Noise level in (dBA)	*Duration in (hr./day)*
80	Any time
85	16
90	8
95	4
100	2
105	1
110	0.5
115	0.25

5.8 NUMERICAL EXAMPLES

Example 1 A single machine produces 100 dB noise during starting the machine. If two identical machines take start from same place and same time, what would be the total noise level?

Solution Let us consider L_1 is the noise level produced by the machine and the corresponding intensity is I_1. I_0 is the reference intensity and has a value of 10^{-12} W/m². So we can write that

$$L_1 = 10 \log \frac{I_1}{I_0} = 100, \quad \log \frac{I_1}{I_0} = 10$$

Or, $I_1 = I_0 10^{10}$

As there are two such machines producing noise, the resultant intensity is given by $I_2 = 2I_1 = 2I_0 10^{10}$ and hence the total loudness is given by

$$L = 10 \log \frac{I_2}{I_0} = 10 \log \frac{2I_0 10^{10}}{I_0}$$

$$= 10(\log 2 + \log 10^{10}) = 10(0.30103 + 10) = 103 \text{ dB}$$

So the total noise level is 103 dB.

Example 2 Estimate the ratio of intensities of two noise levels 90 dB and 100 dB.

Solution Let us consider I_1 and I_2 are the intensities of the noise levels 90 dB and 100 dB respectively. Thus we can write

$$L_1 = 90 = 10 \log \frac{I_1}{I_0} \text{ and } L_2 = 100 = 10 \log \frac{I_2}{I_0}$$

Hence, $I_1 = I_0 10^9$ and $I_2 = I_0 10^{10}$

Thus, $\frac{I_2}{I_1} = 10$

So, the noise level of 100 dB is 10 times louder than the noise level of 90 dB.

Example 3 In a work area the noise levels are read as 95 dB(A) for 2 hr/day, 90 dB(A) for 4 hr/day and 80 dB(A) for 2 hr/day. Determine whether the existing condition is good for health of a worker or not.

Given: Total working hours is 8 hr/day. Noise threshold limit values for 95 dB(A) is 2 hr, 90 dB(A) is 4 hr and 80 dB(A) is 16 hr.

Solution For the noise level of 95 dB(A)

$C_1 = 2$ hr/day and $T_1 = 2$ hr/day

For the noise level of 90 dB(A)

$C_2 = 4$ hr/day and $T_2 = 4$ hr/day

For the noise level of 80 dB(A)

$C_3 = 2$ hr/day and $T_3 = 16$ hr/day

$$\text{So, } \frac{C_1}{T_1} + \frac{C_2}{T_2} + \frac{C_3}{T_3} = \frac{2}{2} + \frac{4}{4} + \frac{2}{16} = 2\frac{1}{8}$$

As the combined effect of noise level exceeds NTLV, i.e., 1, existing environment is not good for the worker.

Example 4 Noise in an area is measured as follows: 95 dB(A) for 1 hr/day, 90 dB(A) for 2 hr/day and 70 dB(A) for 5 hr/day. State whether the existing condition is healthy.

Given: Total working hours is 8 hr/day. Noise threshold limit values for 95 dB(A) is 2 hr, 90 dB(A) is 4 hr and 70 dB(A) is any time.

Solution For the noise level of 95 dB(A)

$C_1 = 1$ hr/day and $T_1 = 2$ hr/day

For the noise level of 90 dB(A)

$C_2 = 2$ hr/day and $T_2 = 4$ hr/day

For the noise level of 70 dB(A)

$C_3 = 5$ hr/day and $T_3 =$ any time/day

$$\text{So, } \frac{C_1}{T_1} + \frac{C_2}{T_2} + \frac{C_3}{T_3} = \frac{1}{2} + \frac{2}{4} + \frac{2}{\infty} = 1$$

As the combined effect of noise level just reaches NTLV, i.e., 1, existing environment is healthy.

Example 5 If a point source develops free progressive waves and the sound pressure level is 85 dB at 6 m from the source, calculate the sound power level of the source.

Solution $L_I = L_p = 85$ dB. Radius of the spherical surface is 6 m. Surface area = $4\pi r^2 = 4\pi 6^2 = 452$ m^2

$L_w = L_I + 10 \log S = 85 + 10 \log 452 = 11.5$ dB.

Addition of two sound levels

(i) Intensity addition $I = I_1 + I_2$.

(ii) Pressure addition $p_{av}^2 = p_1^2 + p_2^2$.

Example 6 If four individual sources have the following sound intensity levels, $A = 90$ dB, $B = 80$ dB, $C = 70$ dB and $D = 60$ dB, find the total sound pressure level (L_p) when all sources operate simultaneously.

Solution

$$90 = 10 \log \frac{p_1^2}{p_0^2}, \qquad 80 = 10 \log \frac{p_2^2}{p_0^2},$$

$$70 = 10 \log \frac{p_3^2}{p_0^2}, \qquad 60 = 10 \log \frac{p_4^2}{p_0^2}$$

$$\frac{p_{av}^2}{p_0^2} = \frac{p_1^2}{p_0^2} + \frac{p_2^2}{p_0^2} + \frac{p_3^2}{p_0^2} + \frac{p_4^2}{p_0^2} \qquad \text{and } L_p = 10 \log \frac{p_{av}^2}{p_0^2} = 90.5 \text{ dB}.$$

Example 7 A person is standing at a distance R from a small source of sound in an open ground and reflections may be neglected. The sound intensity level is L. If the person moves to a distance nR, where $n = 1, 2, 3$ etc., what will be the new sound level?

Solution In an open ground, where there is no reflection, the sound intensity (I_1) is proportional to $1/R^2$, where, R is the distance from the source. If the distance increases from R to nR, intensity decreases from I_1 to I_2 and $I_2 \infty 1/(n^2R^2)$.

$$\text{Thus, } \Delta L = L_2 - L_1 = 10 \log \frac{I_2}{I_0} - 10 \log \frac{I_1}{I_0} = 10 \log \frac{I_2}{I_1} = 10 \log \frac{1}{n^2} = -20 \log n$$

Example 8 Suppose a sound source emits 1 kHz tone at 10 dB. How much must the power of the sound source be increased to raise the sound to 110 dB?

Solution Suppose the power levels of the two signals are P_1 and P_2 respectively. Thus the difference in decibels is given by

$$\Delta L = 10 \log \frac{P_2}{P_1} \text{ dB}. \quad \Delta L = 110 - 10 = 100 \text{ dB}. \quad \frac{P_2}{P_1} = 10^{10}$$

Thus to raise sound intensity from 10 dB to 110 dB, sound power level is increased by 10^{10} times.

Example 9 An amplifier has an input of 2 mV and output of 4 V. What is the voltage gain in dB?

Solution Power dissipated through an electric resistor is proportional to V^2. Thus

$$\text{Voltage gain} = 10 \log \frac{V_{out}^2}{V_{in}^2} = 20 \log \frac{4000}{2} = 20(\log 2 + 3) = 66 \text{ dB}$$

6

Solid Waste Management

6.1 SOIL POLLUTION

6.1.1 Causes of soil pollution

1. Direct discharge of harmful industrial chemicals on land without treatment.
2. Excessive use of pesticides in agriculture causes severe soil pollution.
3. Direct discharge of nuclear wastes and radioactive minerals in ocean and land.
4. Direct discharge of municipal solid waste on the fertile land area.
5. Deforestation creates desert area.
6. Overharvesting.
7. Water pollution.

6.1.2 Adverse effects of soil pollution

6.1.2.1 Effects of industrial pollutants

(a) Industrial wastes consist of variety of chemicals, which are extremely toxic to living being.

(b) Industrial effluents are often discharged to sewage system and hence sewage purification mechanism is adversely affected, causing soil and water-borne diseases.

(c) Industrial effluents contain heavy metal contaminants, which destroy beneficial microorganisms in the soil.

(d) Soluble salts in the effluent cause crop loss and soil corrosion.

(e) Some tannery waste contain pathogenic bacteria.

(f) Chemicals kill the soil organisms which have contribution to the structure and fertility of the soil. Furthermore, they also make the soil toxic.

6.1.2.2 Effects of urban waste products

(a) Urban waste is dumped along highways and other places. These wastes spread several chronic diseases.

(b) The waste accelerates the growth of pathogenic bacteria and viruses.
(c) Solid wastes cause offensive odour and cause clogging of ground water filters.
(d) Deforestation.
(e) Salinity of soil increases, thereby degrading soil quality.

6.1.2.3 Effects of radioactive pollutants

(a) Radiation affects the soil and soil fertility and also kill plant species.
(b) It is responsible for malfunctioning of human organs.
(c) Radio nuclides affect plant growth and destroy crops.

6.1.2.4 Effects of fertilizers

(a) Use of organic fertilizers degrades the soil quality.
(b) Potassium fertilizers will decrease vitamin C and carotene in vegetables and fruits.
(c) Nitrogenous fertilizers lead to accumulation of nitrates in the soil, causing diarrhoea and cyanosis in children. Excessive nitrogen fertilizers produce toxic nitrates and nitrites in different fruits.
(d) Phosphate fertilizers adversely affect the crop yield.
(e) High amount of fluorides in the soil spread fluorosis.

6.1.2.5 Effects of pesticides

(a) Pesticides kill the soil organisms which have contribution to the structure and fertility of the soil. Furthermore, they also make the soil toxic.
(b) Pesticides degrade the quality of crops and vegetables.
(c) It is run off and ultimately mixed with water reservoirs, contaminating drinking water.
(d) Organophosphates pesticides cause extreme muscular weakness, tremors and dizziness.
(e) It also affects tissues and metabolic activities.

6.1.3 Control of soil pollution

(i) Restricted use of chemical fertilizers and pesticides. Use of biofertilizers is recommended.
(ii) Recycling of solid wastes not only helps restoring the soil quality but also conserves the natural resources.
(iii) Plantation of trees to check deforestation.
(iv) Maintenance of soil quality to check soil erosion.
(v) Domestic and municipal wastes should be treated anaerobically to generate biogas and sludge which can be used as fertilizers.
(vi) Limited harvesting.
(vii) Checking water pollution.

6.2 SOIL PROFILE

The nature (thickness, specific gravity, colour, porosity, acidity etc.) and constituents of soil change as we move from the top soil to the core. This gratification of soil along the thickness is called soil profile. So it is the natural change in soil quality across the soil thickness. Lithosphere has three layers: (a) Crust, (b) Mantle and (c) Core.

(a) **Crust** The thickness of the crust is 64–96 km. Main constituents of crust are

(i) *Inorganics* Silicates (O_2, Si and Al) are the major and essential component of soil. Minor components are Na, K, Ca, Mg etc. O : 46%, Si : 27%, Al : 8%, Fe : 5%, others : 1%–3%.

(ii) *Organics* It constitutes 2.5% of the crust. Decomposition of dead plants/ animals by microorganisms produces organics in the soil. The process is called humification and the organics is called humus.

(iii) Water and air.

(iv) Soil bacteria.

The Crust can further be divided into four zones as shown in the Fig. 6.1.

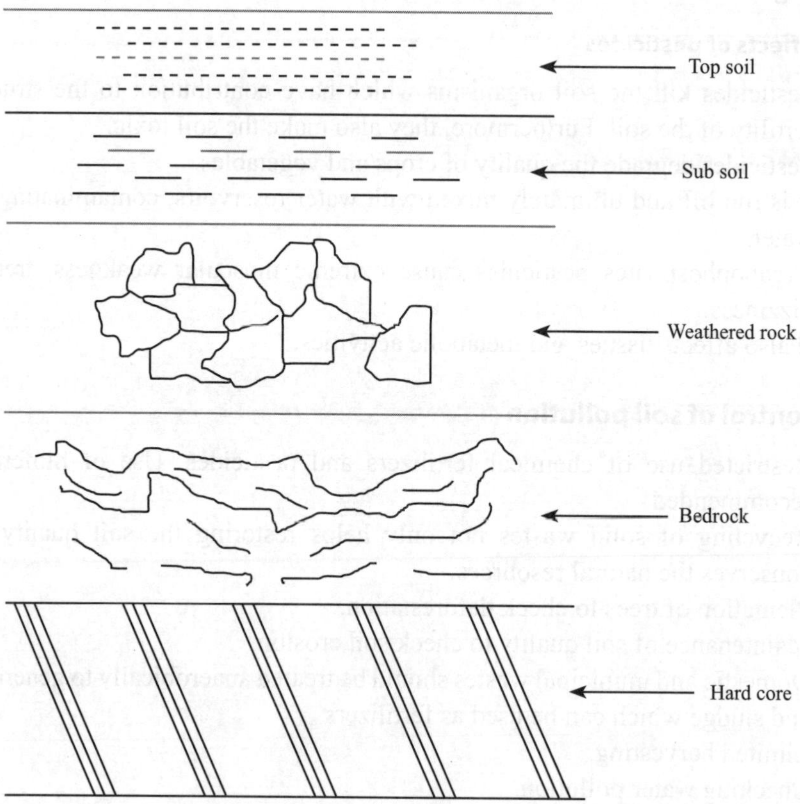

Fig. 6.1 Different zones of Earth crust and Core.

A-horizon: This layer is also called top soil. It contains huge amount of nutrients with water and air, which are essential for plant growth.

B-horizon: This layer is also called sub soil. It contains less amount of nutrients with water and air. Deeper region contains clay materials, iron and organics.

C-horizon: It is the first soil layer, originating from bedrock.

Bedrock: It is the bottom layer of the crust. It is a pure rock but upper soil layer is developed from this rock due to weathering.

(b) **Mantle** The thickness of the mantle is 2800 km. The upper part of the mantle has specific gravity 3.5 and is called upper Mantle. The lower part of the mantle has specific gravity 4.5 and is called lower Mantle. Hard rock is the prime constituent of the mantle.

(c) **Core** The core has an average specific gravity 13. It has two parts. Inner part of the core is solid with thickness 1250 km. The outer part of the core contains molten metals, mostly Fe, Ni and Co. The thickness is 2250 km. Temperature is 5000ºC–6000ºC.

6.3 SOLID WASTE MANAGEMENT

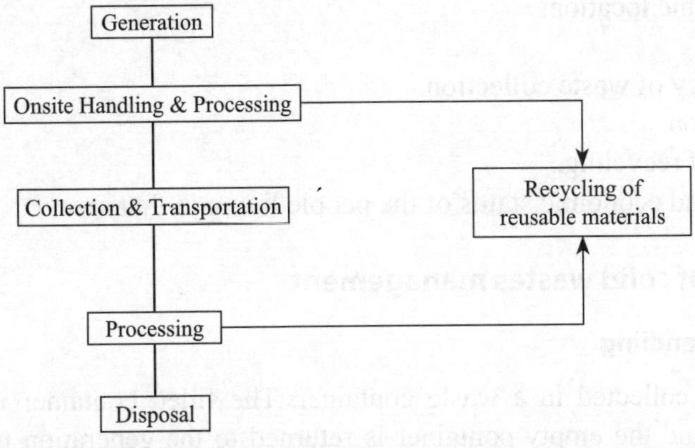

The solid materials generated and discarded due to human and animal activities are called solid wastes. It can be divided into following categories:

(a) Municipal solid wastes, known as MSW.
(b) Biomedical solid wastes, known as BMW.
(c) Industrial solid wastes, known as ISW.
(d) Hazardous solid wastes, known as HSW.

6.3.1 Hazardous solid wastes

Environmental Protection Agency (EPA) has identified four major characteristics of hazardous substance. They are: (a) Ignitability, (b) Reactivity, (c) Corrosiveness, and (d) Toxicity.

So, any substance possessing any one of the above characteristics is called hazardous substance.

Ignitable substance The material, which easily catches fire and continues to burn even if the source of fire is taken away, is called ignitable substance.

Reactive substance The material, which releases toxic fumes and gases when heated, is called reactive substance.

Corrosive substance The liquid, having pH less than 2 or more than 12, is called corrosive liquid.

Toxic substance The material, which has harmful effect on human body whenever ingested or injected or absorbed, is called toxic substance.

6.3.2 Components of MSW

(a) *Garbage* It is the solid food wastes, generated from house.
(b) *Rubbish* It is the household solid wastes and the wastes generated due to construction and demolition.
(c) *Ashes and residues* It is powdery materials and burnt materials.
(d) *Other wastes* Street sweeping, dead animals, debris etc.

6.3.2.1 Factors influencing solid wastes generation

(a) Geographic location.
(b) Season.
(c) Frequency of waste collection.
(d) Population.
(e) Extent of recycling.
(f) Habits and economic status of the people living in that area.

6.3.3 Stages of solid wastes management

6.3.3.1 Onsite handling

Solid wastes are collected in a waste container. The filled container is emptied at the collection point and the empty container is returned to the generation point for the next collection. Factors influencing onsite handling are

(i) Type of solid waste.
(ii) Frequency of collection.
(iii) Capacity of container.

6.3.3.2 Onsite processing

It is the process by which reusable solid materials are recovered from solid wastes.

Collection & transportation

It includes all the activities related to the gathering of solid wastes from the points of generation and transported to the location where collection vehicles are emptied.

6.3.3.3 Processing of solid wastes

Purpose

(a) To improve the efficiency of solid waste management system.
(b) To recover the usable materials for reuse.

Techniques

(a) *Compaction* Reduction in volume is an important factor in solid waste management system. Solid wastes are usually compacted by using special equipment and at high pressure.
(b) *Incineration* In this method solid wastes are burnt in open atmosphere to reduce the volume. Original volume of solid wastes may be reduced by about 80% to 90%. However, air pollution is the main disadvantage of this method.
(c) *Shredding* In this method the size of a given waste is reduced almost uniformly by grinding.
(d) *Component separation* It is a process by which reusable materials are recovered from solid wastes. It can be done either manually or mechanically. In the manual separation process further processing of wastes is not required. However, in mechanical separation process, magnetic separation, screening etc. are used to separate reusable materials.
(e) *Drying* Sludge materials of different treatment plants are first compacted, filtered and then dried up before extracting reusable materials.

6.3.3.4 Disposal of solid wastes

(a) Open dumping

Generally low lying areas of the outskirts of cities are used for that purpose. However, the open dumping of solid wastes pollutes the environment significantly and damages public health.

(b) Incineration

Simple burning of solid wastes at high temperature is called incineration. However, it leads to sufficient air pollution. So the incineration plant should be well designed as per air pollution standards.

(c) Sanitary land filling

Solid waste is sprayed and compacted in thin layers within a small area. This layer is called cell. The cell depth usually lies within 2 m. The cell is sealed by a cover of thickness 20 cm. Several of such cells are placed uniformly followed by compaction. The method is biological treatment of solid wastes. The treatment procedure has the following stages

- **1st phase** Aerobic decomposition is allowed to take place with increase in oxygen deficit along with the liberation of heat. So temperature increases.

- **2nd phase** In this phase oxygen deficit reaches maximum and anaerobic decomposition occurs with evolution of hydrogen, methane, hydrogen sulphide, ammonia etc. Methane gas is produced continuously and ultimately the waste is stabilized, indicating the end of the 2nd phase.
- **3rd phase** Anaerobic decomposition is completed and aerobic condition prevails.

A sanitary landfill system is shown in the Fig. 6.2.

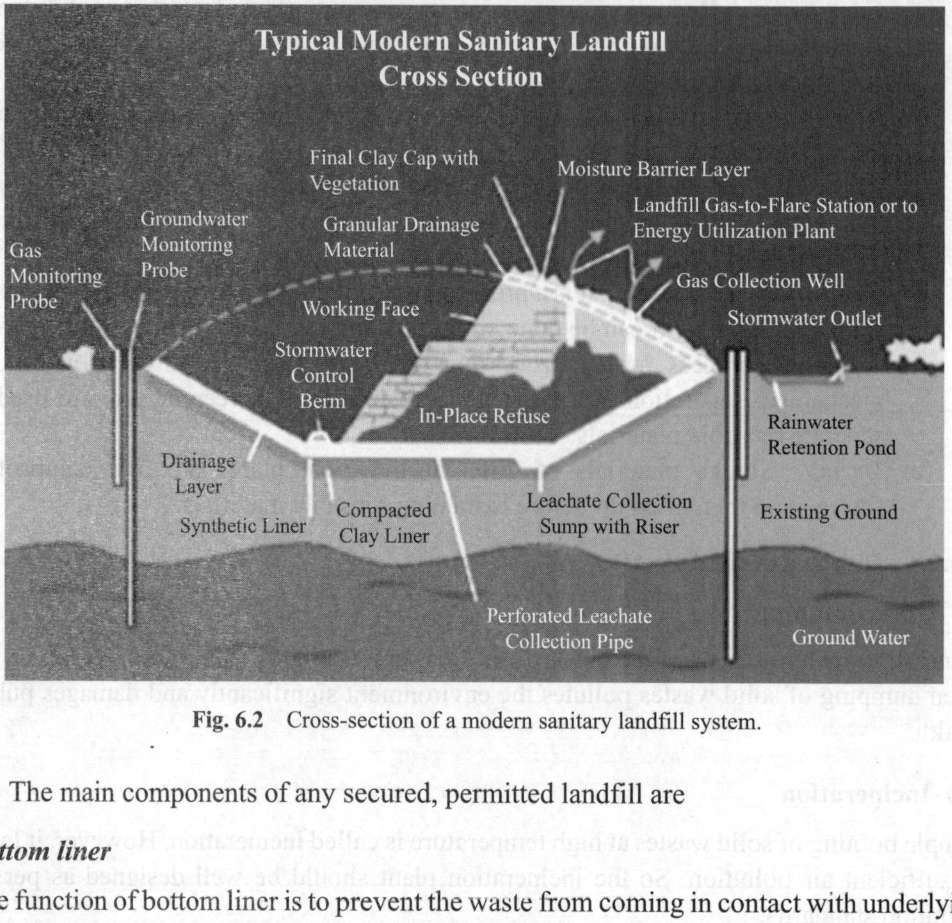

Fig. 6.2 Cross-section of a modern sanitary landfill system.

The main components of any secured, permitted landfill are

Bottom liner

The function of bottom liner is to prevent the waste from coming in contact with underlying natural soils and groundwater. It is made of a composite material (combination of clay soil and HDPE (High Density Polyethylene) plastic film of thickness ranging from 1 mm to 2.5 mm).

Cells (old and new)

The solid waste is disposed in these cells. Cells cover a wide area as large as 20 *acres*. Each cell is again divided into different small cells. Here the waste is compacted and shredded by heavy landfill compaction machinery.

Leachate collection system

The bottom surface of each cell is sloped to a slump, where trapped liquid in solid waste is collected. This is called leachate. The leachate collection system consists of a series of perforated pipes, gravel packs and a layer of sand placed in the bottom of the landfill. Finally the leachate is pumped out and stored in a pond, where it is treated before discharging to a reservoir.

Methane collection system

The waste in the cell is treated with bacteria in absence of oxygen, producing huge amount of methane. As methane gas often explodes, it is removed from the landfill with the help of a series of pipes and collected.

Cover (or cap)

The cell is to be covered with at least 6 inches compacted soil in order to avoid spreading foul odours and to isolate the waste from exposure to air and animals, such as birds, rats, mice etc. When the cell is completely filled with waste, it is permanently covered with HDPE layer, compacted soil and topsoil. The topsoil helps growth of crops and also prevents erosion of soil.

Groundwater monitoring stations

A series of wells are to be set up surrounding the cells. Some wells are set up up-gradient with respect to cells and other wells are set up down-gradient with respect to cells. Water quality of both type of wells is continuously monitored in order to avoid any type of contamination of the groundwater.

(d) Composting

Microorganisms, present in the solid wastes, decompose the organics to produce soil conditioner. The method is known as composting. It is an aerobic process. Initially temperature increases and then decreases. This decrease in temperature indicates the end point of composting process.

- **1st stage** Mesophilic bacteria decomposes organics to release CO_2 and heat. Temperature rises to about 45°C.
- **2nd stage** Thermophilic bacteria continues decomposition process with rise in temperature. Temperature rises to about 60°C.
- **3rd stage** Temperature decreases, indicating that the decomposition process is complete and the waste becomes stabilized. The final compost has a typical smell and dark brown colour.

7

Ozone

7.1 INTRODUCTION

Ozone is available in both troposphere and stratosphere. However, most of the ozone is concentrated in stratosphere. The ratio of concentration of oxygen to ozone, i.e., $[O_2]/[O_3]$ is around 10^{25} in the troposphere but it is around 3×10^5 in the stratosphere.

Ozone is a colourless, highly reactive gas with a distinctive odour. It is formed usually in the upper atmosphere (the stratosphere) and protects the Earth from ultraviolet (UV) radiation from the Sun. However, at ground level, sufficient concentration of ozone is produced through photochemical reactions involving other air pollutants, such as, nitrogen dioxide (NO_2), oxygen (O_2) and hydrocarbons. That is why ozone is called **secondary air pollutant** in the troposphere. Tropospheric ozone has harmful effect on human health but stratospheric ozone protects life on Earth. Hence stratospheric ozone is often called "good ozone" and tropospheric ozone is called "bad ozone".

7.2 UV RADIATION AND ITS HARMFUL EFFECT

The spectrum of light wave is shown in the following figure.

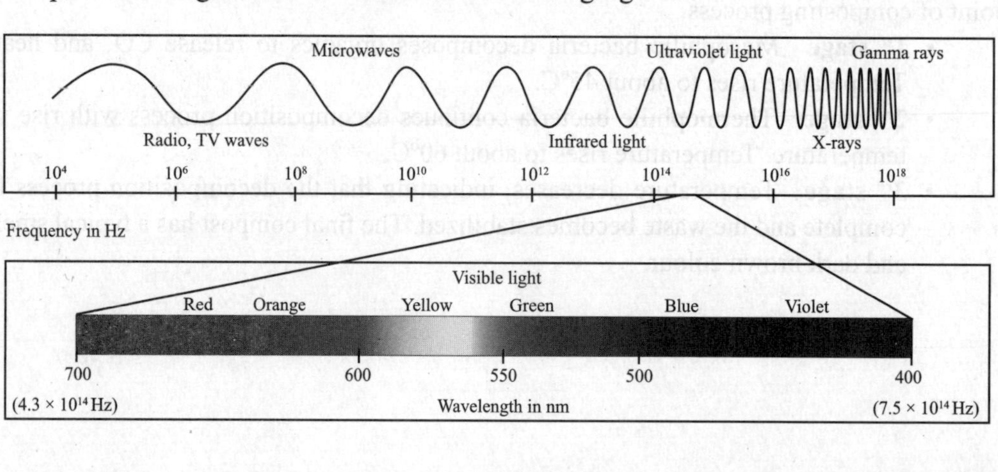

Ultraviolet radiation is one form of radiant energy coming from the Sun. The various forms of energy or radiation are classified according to wavelength, measured in nanometres (one nm is a millionth of a millimetre). The shorter the wavelength, the more energetic the radiation is. In order of decreasing energy, the principal forms of radiation are gamma rays, X-rays, UV (ultraviolet radiation), visible light, infrared radiation, microwaves and radio waves. Broadly, there are three categories of UV radiations

- UV-A, wavelength range 320-400 nm
- UV-B, wavelength range 280-320 nm
- UV-C, wavelength range 200-280 nm

Generally, the shorter the wavelength, the more biologically damaging UV radiation can be if it reaches the Earth in sufficient quantity. UV-A is the least damaging (longest wavelength) form of UV radiation and reaches the Earth in greatest quantity. Most UV-A rays pass right through the ozone layer in the stratosphere. UV-B radiation can be very harmful. Fortunately, most of the Sun's UV-B radiation is absorbed by ozone in the stratosphere. UV-C radiation is potentially the most damaging because it is very energetic. Fortunately, all UV-C is absorbed by oxygen and ozone in the stratosphere and never reaches the Earth's surface. In summary, the danger from ultraviolet radiation comes mainly from the UV-B range of the spectrum, although UV-A poses some risk if exposure is long enough, or the sunshine is particularly strong.

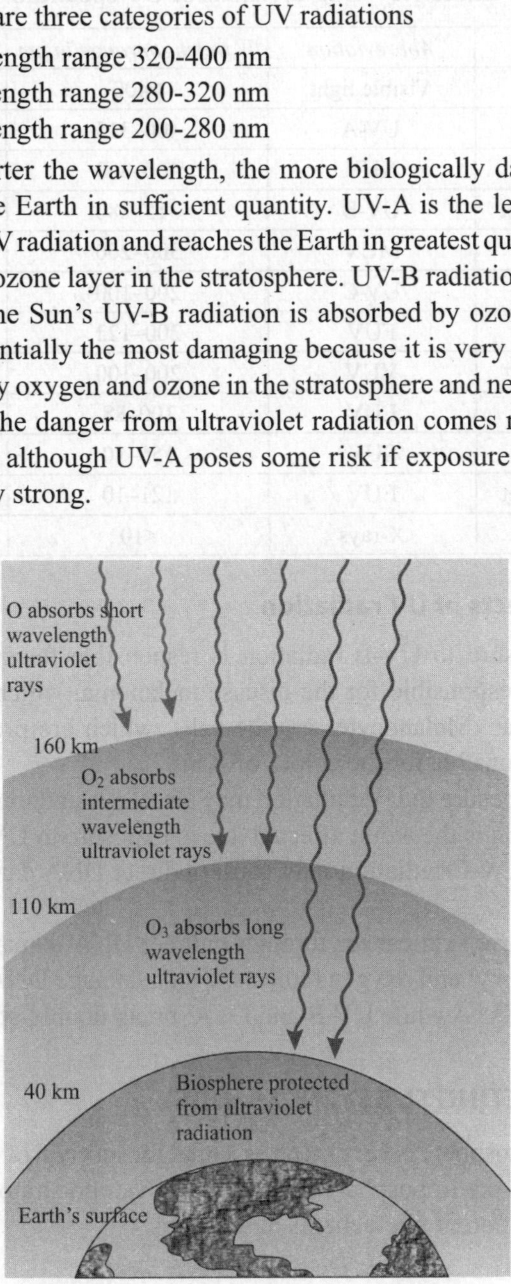

The electromagnetic spectrum of ultraviolet light can be subdivided in a number of ways. According to ISO (ISO-DIS-21348) the detailed divisions of UV-spectrum alongwith their wavelength and energy are described in the following table

Table 7.1 Sub-divisions of UV-spectrum.

Name	*Abbreviation*	*Wavelength range in nm*	*Energy per photon in ev*
Before UV spectrum	Visible light	>400	<3.1
Long wave Ultraviolet	UV-A	400–315	3.10–3.94
Near Ultraviolet wave	NUV	400–300	3.10–4.13
Medium wave Ultraviolet	UV-B	315–280	3.94–4.43
Middle wave Ultraviolet	MUV	300–200	4.13–6.20
Short wave Ultraviolet	UV-C	200–100	4.43–12.4
Far wave Ultraviolet	FUV	200–122	6.20–10.20
Vacuum wave Ultraviolet	VUV	200–100	6.20–12.40
Low wave Ultraviolet	LUV	100–88	12.40–14.10
Super wave Ultraviolet	SUV	150–10	8.28–124
Extreme wave Ultraviolet	EUV	121–10	10.2–124
Beyond UV rays	X-rays	<10	>124

7.2.1 Harmful effects of UV radiation

- An overexposure to UV-B radiation is responsible for sunburn and some skin cancer. It is responsible for the disease melanoma, which is a malignant tumor of melanocytes. Melanocytes are the cells, which are producing dark pigment, melanin, responsible for the colour of skin.
- Prolonged exposure to UV radiation may affect eye and immune system of human being. The skin is the worst affected due to exposure to UV radiation.
- Exposure to UV-C radiation may cause indirect DNA damage and carcinogenic diseases.
- UV-A can cause skin cancer through indirect DNA damage. It generates highly reactive hydroxyl and oxygen radicals, which damage DNA. UV-A mostly breaks single-strand DNA while UV-B and UV-C break double-strand DNA.

7.3 OZONE DEPLETING SUBSTANCES (ODS)

The ozone layer in stratosphere is very much essential for survival of life on Earth. However, pollutants, available in the troposphere, frequently move into stratosphere and destroy the ozone layer through different photochemical reactions.

7.3.1 Introduction

CFCs are not the only ozone-depleting chemicals (ODCs). A number of other halocarbon species are capable of destroying ozone in the stratosphere. Halocarbons include the chlorofluorocarbons (CFCs), the hydrochlorofluorocarbons (HCFCs), methyl halides, carbon tetrachloride (CCl_4), carbon tetrafluoride (CF_4) and the halons (bromide species). Although the HCFCs do not contribute significantly to the destruction of the ozone layer, they, along with the other halocarbons, are all considered to be powerful greenhouse gases and have significant contribution to global warming.

7.3.2 Hydrofluorocarbons

Hydrochlorofluorocarbons or HCFCs contain chlorine but, like CFCs, they also contain hydrogen (the *H*) which causes them to break down in the lower atmosphere (troposphere). They are called transition chemicals because they are considered an interim step between strong ozone-depleters and replacement chemicals that are entirely ozone-friendly. Unfortunately, like CFCs, they are strong greenhouse gases and contribute towards global warming. Some examples of ODS are given below.

7.3.2.1 Carbon tetrachloride

Carbon tetrachloride (CCl_4), despite its toxicity, was first used in the early 1900s as a fire extinguisher, and more recently as an industrial solvent, an agricultural fumigant and in many other industrial processes including petrochemical refining, and pesticide and pharmaceuticals production. Recently it has also been used in the production of CFC-11 and CFC-12. It has accounted for less than 8% of total ozone depletion. The use of carbon tetrachloride in developed countries has been prohibited since the beginning of 1996 under the Montreal Protocol agreement.

7.3.2.2 Methyl chloroform

Methyl chloroform, also known as 1, 1, 1-trichloroethane is a versatile, all-purpose industrial solvent used primarily to clean metal and electronic parts. It was introduced in the 1950s as a substitute for carbon tetrachloride. Methyl chloroform has accounted for roughly 5% of total ozone depletion. The use of methyl chloroform in developed countries has been prohibited since the beginning of 1996 under the Montreal Protocol agreement.

7.3.2.3 Halons

Halons, unlike CFCs, contain bromine, which also destroys ozone in the stratosphere. Halons are used primarily as fire suppressants. Halon-1301 has an ozone depleting potential 10 times that of CFC-11. Although the use of halons in developed countries has been phased out since 1996, the atmospheric concentration of these potent, long-lived ozone destroyers is still rising by an estimated 11 to 15% annually. To date halons have accounted for about 5% of global ozone depletion.

7.3.2.4 Methyl bromide

Methyl bromide, another bromine-containing halocarbon, has been used as a pesticide since the 1960s. Today, scientists estimate that human sources of methyl bromide are responsible for approximately 5 to 10% of global ozone depletion.

7.3.3 CFC numbering

The molecular formula of CFC can be identified from the CFC number. Conversely, CFC-number can be identified from the knowledge of structure. Some examples are given below

(i) *Find the structure of CFC-115.* At first 90 should be added to 115 and the result is 115 + 90 = 205. The 1st figure, i.e., 2 indicates number of C-atoms, the 2nd figure, i.e., 0 indicates number of H-atoms and the 3rd figure, i.e., 5 indicates number of F-atoms. All other atoms, attached to C-atoms, are Cl-atoms. Here only one Cl-atom is needed to satisfy the valencies of two C-atoms. Thus the molecular formula of CFC-115 is C_2F_5Cl.

(ii) *Find the structure of CFC-12.* 12 + 90 = 102. So, the 1st figure, i.e., 1 indicates number of C-atoms, the 2nd figure, i.e., 0 indicates the number of H-atoms and the 3rd figure, i.e., 2 indicates the number of F-atoms. So, 2 Cl-atoms are needed to satisfy the valency of one C-atom. Thus the molecular formula of CFC-12 is CF_2Cl_2.

(iii) *Find the CFC number of CHF_2Cl.* Number of carbon atoms = 1, Number of hydrogen atoms = 1 and number of fluorine atoms = 2. So the number is 112 – 90 = 22. The compound is CFC-22.

(iv) *Find the CFC number of $C_2H_2F_3Cl$.* Number of carbon atoms = 2, Number of hydrogen atoms = 2 and number of fluorine atoms = 3. So the number is 223 – 90 = 133. The compound is CFC-133.

7.3.4 Applications of ODS

(a) CFCs and HFCs are widely used as propellants.

(b) CFCs and HFCs are also used as refrigerants in air conditioners, refrigerators, chillers etc.

(c) Halons have unique characteristic to arrest flame propagation. So they are widely used to extinguish fires.

(d) CFCs and HFCs act as blowing agents. Hence, CFCs and HFCs are used to manufacture foams.

(e) Carbon tetrachloride, methyl chloroform, CFCs and HFCs are widely used as cleaning agents for cleaning printed circuit boards, and also for degreasing different metal parts.

(f) Carbon tetrachloride, methyl chloroform, CFCs and HFCs are also used in inks, coatings and medical applications.

(g) Methyl bromide is frequently used as fumigant.

(h) Carbon tetrachloride is used as (i) raw materials for the production of CFC-11 and CFC-12, (ii) processing agent for some pharmaceuticals and agricultural chemicals, (iii) processing agent in the production of chlorinated rubber, (iv) laboratory reagent for different research activities.

(i) A gas mixture containing 88% CFC-12 and 12% ethylene oxide (EO) is used as sterilant in different pharmaceutical laboratories and hospitals to sterilize different medical equipment.

(j) Methyl chloroform, which is nonflammable and quick drying agent, is extensively used in adhesive manufacturing industries.

(k) HFC-134a is widely used as a refrigerant fluid in different automobiles.

(l) HFC-22 is used as refrigerant fluid in small water chillers while CFC-11 and CFC-12 are used in large water chillers. HCFC-22 and HFC blends are used in cold storage and food processing.

7.4 MEASURING OF OZONE LAYER THICKNESS

The most common ozone measurement unit is the Dobson Unit (DU). The scientist G.M.B. Dobson carried out extensive research work on ozone in the atmosphere for more than 40 years in around 1920–1970. He successfully established a method to measure ozone layer thickness in the atmosphere. The unit of ozone layer thickness is expressed in Dobson Unit or DU after his name. A DU measures the total amount of ozone in an overhead column of the atmosphere. If the total amount of ozone in the overhead column of the atmosphere of an area were compressed into one layer at 0 degree Celsius and a pressure of one atmosphere, the height of the column indicates the thickness of the ozone layer in that area. Every 0.01 millimetre thickness of the layer is equal to one Dobson Unit. The average amount of ozone in the stratosphere across the globe is about 300 DU (or a thickness of only 3 mm at 0ºC and 1 atmospheric pressure). The highest ozone layer thickness is found in the mid to high latitudes in Canada and Siberia (360 DU).

7.4.1 Formation of ozone hole

Chlorine and bromine, remaining in the atmosphere as stable compounds, are inactive. So they do not destroy the ozone. However, during the Antarctic winter months (June to August) when the region receives no sunlight, the stratosphere becomes cold enough (–80ºC) to form high thickness ice cloud, called Polar Stratospheric Clouds (PSCs). These PSCs provide an ideal catalytic surface on which chlorine can react with the ozone, destroying the ozone layer. As this reaction requires sunlight, the reaction sets in only when the Sun returns to Antarctica in spring (September to October), before the PSCs have had a chance to melt. During late spring and summer the Antarctic air sufficiently warms up to melt the ice clouds and hence ozone hole disappears. During the southern hemisphere winter season, Antarctica is isolated from the rest of the world by a natural circulation of wind called the polar vortex. This prevents atmospheric mixing of stratospheric ozone, thus contributing to the depletion of ozone.

7.4.2 Beneficial effects of ozone

Ozone in stratosphere region has the following beneficial effects on all living beings

(i) Ozone layer acts as a protective shield for living organisms on Earth from various injurious effects of UV rays (220–230 nm), which is proven to be responsible for sunburn and cancer.

(ii) In stratosphere ozone layer supplies necessary heat through absorption of UV rays to develop a positive lapse rate in the stratosphere due to which stratospheric region is quite stable and pollutants, if enters into the stratosphere, is entrapped locally instead of diffusion.

7.5 ENEMIES OF OZONE

Prime enemies of stratospheric ozone are: CFC, NO, OH^*, Cl_2 and Pb.

In presence of CFC the following chain reaction is believed to take place.

$$CFCl_3 \xrightarrow{hv} CFCl_2^* + Cl^*$$
$$Cl^* + O_3 \longrightarrow ClO^* + O_2$$
$$ClO^* + O \longrightarrow Cl^* + O_2$$

Chain reaction continues.

However, NO_2, present in the atmosphere, deactivates ClO^* by forming chlorine nitrate and hence helps reduce ozone depletion. The reaction is shown below

$$ClO^* + NO_2 \longrightarrow \underset{\text{Chlorine nitrate}}{ClONO_2}$$

In presence of NO the following reactions may occur.

$$NO + O_3 \longrightarrow NO_2 + O_2$$
$$NO_2 + O_3 \longrightarrow NO_3 + O_2$$
$$NO_2 + NO_3 \longrightarrow N_2O_5$$
$$N_2O_5 + H_2O \longrightarrow 2HNO_3$$

HNO_3 is precipitated as in the form of acid rain on the Earth's surface.

In presence of Cl_2 the following reactions may occur.

$$Cl_2 \xrightarrow{hv} 2Cl^*$$
$$Cl^* + O_3 \longrightarrow ClO^* + O_2$$
$$ClO^* + O \longrightarrow Cl^* + O_2$$
$$Cl^* + RCH_3 \longrightarrow RCH_2^* + HCl$$

HCl is precipitated as acid rain on the Earth's surface.

TEL (tetraethyl lead) is the primary source of OH^*, which then reacts with stratospheric ozone to form peroxide radical.

$$Pb(C_2H_5)_4 + O_2 \longrightarrow PbO_2 + 4C_2H_5^*$$

$$PbO_2 + RCH_2^* \longrightarrow PbO + RCH_2O^*$$
$$RCH_2O^* + O_2 \longrightarrow RCHO + HOO^*$$
$$HOO^* + PbO \longrightarrow PbO_2 + HO^*$$
$$OH^* + O_3 \longrightarrow O_2 + HOO^*$$

Here $C_2H_5^*$ can be written as RCH_2^*. So there is no problem.

7.6 OZONE DEPLETING POTENTIAL

The Scientist, Wuebbles in 1983 first introduced the term ozone depleting potential or ODP. He defined it as a measure of destructive effects of an ODS with respect to a reference substance. It refers to the amount of a given ODS required to destroy 0.01 mm ozone layer thickness with respect to that of CFC-11. Among the chlorofluorocarbons CFC-11 has the maximum potential to destroy ozone layer thickness since it has three chlorine atoms in the molecule. Thus CFC-11 has been taken as reference and its ODP is 1. In case of chlorofluorocarbons ODP is nearly equal to 1, brominated substances have ODP in the range of 5–15. ODP of hydrofluorocarbons (HFC) is taken as zero since HFCs have no chlorine or bromine atoms. In presence of UV rays, CFCs and halons break to form chlorine and bromine free radicals, which are responsible for fast destruction of ozone layer. One chlorine radical can break more than 10^5 ozone molecules while one bromine radical can eat up more than 4×10^6 ozone molecules. So ODP of an ODS can be written as

$$\text{ODP} = \frac{\text{Ozone layer thickness (in DU) destroyed by a unit mass of ODS}}{\text{Ozone layer thickness (DU) destroyed by a unit mass of CFC-11}}$$

7.7 MONTREAL PROTOCOL

An international agreement was made in 1987 at a conference in Montreal and signed by 34 countries. According to this agreement all countries agreed to ban CFC so that release of CFC in atmosphere would be reduced to 50% by the end of 20th century. Even after enforcement of Montreal Protocol in different countries, it was observed that ozone hole continued to widen at a faster rate. In June 1990 at London Conference the developed countries agreed to ban CFC totally by 2000 A.D. The London amendment of the Montreal Protocol was ratified in May, 1992 and came into effect on 10th August 1992.

7.8 TROPOSPHERIC OZONE

7.8.1 Sources of tropospheric ozone

(i) Emission of motor vehicles, combustion of fossil fuel, industrial and domestic use of solvents and coatings are the major source of ozone.

(ii) The nuclear waste is another important source of ozone.

(iii) Leakage of high voltage electricity from high tension cables.

7.8.2 Adverse effects of tropospheric ozone

(i) Tropospheric ozone causes eye irritation, nose and throat irritations, and breathing difficulties, leading to cardiovascular and respiratory diseases.
(ii) Exposure to 50 *ppm* ozone for several hours will lead to death due to accumulation of fluid in the lungs, leading to severe damage to lung capillaries. Younger people are susceptible to more toxic effects than older ones.
(iii) Ozone is responsible for malfunctioning of heart.
(iv) It damages trees and reduces growth rate.
(v) It reduces crop-yield, such as corn, wheat etc.

7.8.3 Permissible concentration of ozone in air

The average permissible concentration of ozone in air is less than 150 $\mu g/m^3$ for one hour and less than 100 $\mu g/m^3$ for eight hours.

8

Air Pollutants

8.1 INTRODUCTION

There are many species present in the atmosphere with concentration higher than their TLV limits. These are called air pollutants. These pollutants are of three types: (i) primary air pollutants, (ii) secondary air pollutants and (iii) contaminants. However, these air pollutants can be classified in other ways.

8.2 CLASSIFICATION

Air pollutants are broadly classified into six categories.

(a) *Suspended Particulate Matter (SPM)* Pollutants, remaining suspended in air in particulate form, are called suspended particulate matter or SPM. SPM can further be categorized as (i) PM_{10} particles, (ii) $PM_{2.5}$ particles, (iii) aerosols, (iv) soot particles.

(b) *Gaseous pollutants* Mainly three different types of gaseous pollutants are profusely available in the atmosphere, namely, SO_x, NO_x, and CO.

(c) *Smog* It can be divided into two categories: (i) Photochemical smog or oxidising smog, (ii) Sulphurous smog or reducing smog.

(d) *Acid rain* It is a weak acid solution, containing HCl, H_2SO_4 etc.

(e) *Hydrocarbons* It contains mostly gaseous aliphatic and aromatic hydrocarbons.

(f) *Heavy metals* Metals available in nature are harmful to human being. Some metals become harmful only at high concentration. TLV of those metals are high. On the other hand, TLV of a few metals is very low. These are extremely harmful. Examples are As, Pb, Cd and Hg. Specific gravity of these metals is greater than 4. These four metals are collectively called heavy metals.

8.3 SUSPENDED PARTICULATE MATTER (SPM)

Pollutants remain suspended as particles in the atmosphere. These particles are called SPM. Depending on size and nature of particles, SPM can be classified as under

PM_{10} particles Suspended particles, having diameter ≤ 10 μ, are called PM_{10} particles. These particles can remain in the air for several days and can be spread by winds over long distances from the original source. These particles may enter into human body through nose and finally deposited at lungs, causing breathing trouble. However, they have narrow penetration effect.

$PM_{2.5}$ particles Suspended particles, having diameter $\leq 2.5\ \mu$, are called $PM_{2.5}$ particles. These particles can remain in the air indefinitely. These particles have deeper penetration effect and cause severe breathing trouble.

8.3.1 Sources of fine particles

(i) Motor vehicles exhausts are the main source of fine particles.
(ii) Forest fires, domestic fires and combustion of fuels are the other major sources of fine particles.
(iii) Natural hazards, like volcano eruption, blizzards, tornado etc. are also sources of fine particles.
(iv) Fertilizers, pesticides etc., used in agricultural land, also generate SPM.
(v) Fly ash released from power plants and particulate-exhaust from chemical plants.
(vi) Asbestos and marble factories.
(vii) Coal mines, mineral pits etc.

8.3.2 Adverse effects of fine particles

(i) Fine particles carry poisonous carcinogenic (cancer causing) material into the lungs. The finer $PM_{2.5}$ particles have deeper penetration effect into the lungs and are responsible for respiratory diseases and premature deaths. Particularly elderly people and children suffer from respiratory diseases, such as asthma and cardiovascular diseases, due to presence of finer $PM_{2.5}$ particles.
(ii) Fine particles reduce visibility, affect plant growth, thereby, responsible for soil erosion and damaging of minerals.
(iii) Fine particles are believed to provide catalytic surface for different photochemical reactions to generate oxidising and reducing smog, fog, blizzards etc.
(iv) These particles are responsible for unusual electrical phenomena occurring in the atmosphere and climatic change.
(v) SPM damages buildings and sculptures.

8.3.3 Permissible concentration of fine particles in air

The average concentration of SPM should be less than 50 $\mu g/m^3$ per day and less than 20 $\mu g/m^3$ per year.

8.3.4 Aerosol

Small solid particles, dispersed in liquid droplets, is collectively termed particulates. Some colloidal size floating particles, having diameter range 0.1μ–1μ, in the atmosphere, are called aerosols.

8.3.4.1 Sources of aerosol

(i) Fog, bacteria, volcano eruption etc.

(ii) In coal mines and iron-pits FeS_2 is converted to Fe_3O_4 according to the following reactions. These Fe_3O_4 particles are the important constituent of aerosol.

$$3FeS_2 + 8O_2 \longrightarrow Fe_3O_4 + 6SO_2$$

(iii) During combustion of fuel oil, vanadium is converted to V_2O_5, which is also a constituent of aerosol.

(iv) Calcium carbonate ($CaCO_3$), present in coal ash, is converted to solid CaO particles on heating. $CaCO_3 \longrightarrow CaO + CO_2$.

(v) Tetraethyl lead (TEL), present in the gasoline, reacts with halogen scavengers (ethylene dichloride and ethylene dibromide) in presence of oxygen to form halogeneted lead particles.

$$Pb(C_2H_5)_4 + O_2 + C_2H_4Cl_2 + C_2H_4Br_2 \rightarrow CO_2 + H_2O + PbCl_2 + PbBr_2 + Pb(Cl)(Br)$$

(vi) Presence of H_2SO_4 droplets in the atmosphere produces aerosol mists in the environment.

$$2SO_2 + O_2 + 2H_2O \longrightarrow 2H_2SO_4 \text{ (droplets)}$$

$$H_2SO_4 \text{ (droplets)} + 2NH_3 \text{ (gas)} \longrightarrow (NH_4)_2SO_4 \text{ (droplets)}$$

$$H_2SO_4 \text{ (droplets)} + CaO \text{ (droplets)} \longrightarrow CaSO_4 \text{ (droplets)} + H_2O$$

These droplets lose water at low humidity to form solid aerosol.

8.3.5 Polycyclic Aromatic Hydrocarbons (PAH)

8.3.5.1 Introduction

Polycyclic aromatic hydrocarbons or PAH is formed while burning of fossil fuel. Power plant exhaust and automobile exhaust contain sufficient amount of PAH. It appears as soot particles, which is composed of several thousands of crystallites, each of which contains at least 100 aromatic rings. Some examples are given below.

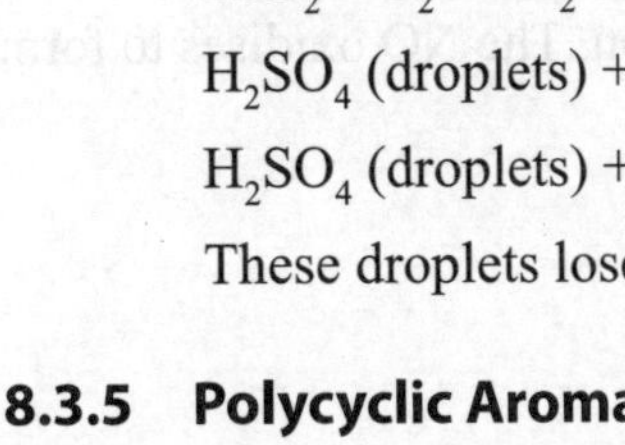

Benzo-(α)-pyrene

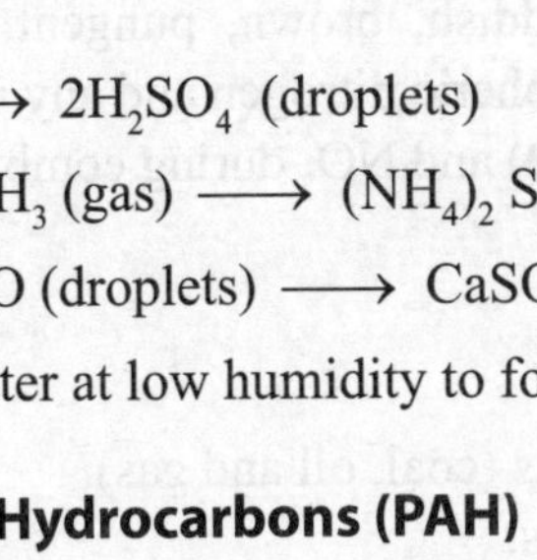

Chrysene

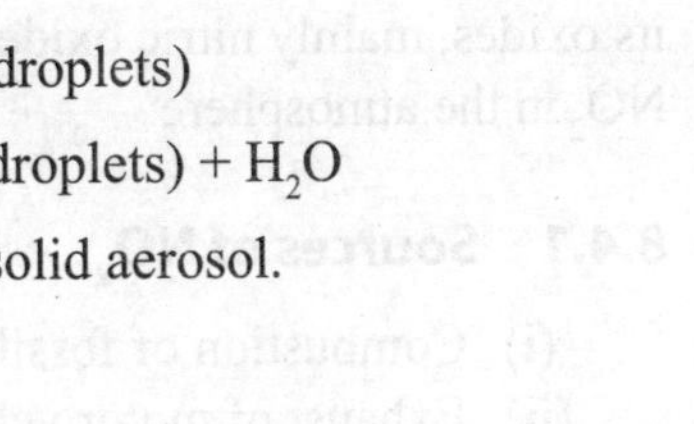

Benzofluoranthane

8.3.5.2 Sources

(i) Bacterial decomposition of organic matters.
(ii) Burning of papers, plastic and rubber.
(iii) Burning of resins.
(iv) Smoking cigarettes.
(v) Automobile exhausts.
(vi) Asbestos factories.

8.3.5.3 Adverse effects

Soot particle provides good catalytic surface to execute different photochemical reactions. It has wide surface area and hence harmful pollutants and metals are adsorbed easily on soot particle surface. It carries carcinogenic chemicals, metals and other harmful pollutants into the human body.

8.4 NITROGEN DIOXIDE (NO_2)

Nitrogen dioxide (NO_2) is a reddish, brown, pungent, acidic gas that is formed by combustion of fossil fuels. Atmospheric nitrogen and any nitrogen in fuel is converted into its oxides, mainly nitric oxide (NO) and NO_2 during combustion. The NO oxidises to form NO_2 in the atmosphere.

8.4.1 Sources of NO_2

(i) Combustion of fossil fuels (coal, oil and gas).
(ii) Exhaust of motor vehicles.
(iii) Combustion of natural gas, refining of petrol and metals.
(iv) Commercial manufacturing and food manufacturing units.
(v) Electricity generation using fossil fuels also produces significant amount of NO_2.
(vi) Bacterial decomposition of organic matter.
(vii) Fertilisers used in agriculture.

8.4.2 Adverse effects of nitrogen dioxide

(i) Nitrogen dioxide damages the lungs, increase severity of asthma and decreases resistance to infections. Children are readily susceptible to respiratory infection.
(ii) Nitrogen dioxide is an important component of photochemical smog, which is harmful to plants, animals and human beings.
(iii) Nitrogen dioxide is responsible for premature leaf fall. Plant growth is suppressed in presence of nitrogen dioxide and hence crop yield is reduced.
(iv) Textile strength and colour fastness of garments are sufficiently deteriorated in presence of nitrogen dioxide.
(v) Nitrogen dioxide undergoes photochemical reaction to form nitric acid, which is precipitated on the Earth in the form of acid rain.

8.4.3 Permissible concentration of nitrogen dioxide in air

The maximum admissible average NO_2 concentration is 200 μg/m³ for one hour and 100 μg/m³ for 24 hours.

8.5 CARBON MONOXIDE (CO)

Carbon monoxide is a colourless, odourless gas formed when substances containing carbon (fuels such as petrol, gas, coal and wood) are burnt with a low supply of air.

8.5.1 Sources and contributors of carbon monoxide

(i) Natural hazards, like volcanic activity and human activities.
(ii) Exhaust of motor vehicles is the other major source of CO.
(iii) Combustion of fossil fuels.
(iv) Disposal of solid waste and land pollution.
(v) Coal mines.
(vi) Power plant exhausts.

8.5.2 Adverse effects of carbon monoxide

(i) CO reacts with oxygen carrying haemoglobin and replaces oxygen in haemoglobin, thereby decreasing the blood's ability to carry oxygen, causing oxygen deficiency at lungs, heart and brain and hence also affects central nervous system. The following reaction is believed to take place
$CO + Hb(O_2) \longrightarrow Hb(CO) + O_2$.
(ii) High level of CO can affect people with heart conditions such as angina, and can impair coordination and attention.
(iii) CO is a known asphyxiant.

8.5.3 Permissible concentration of carbon monoxide in air

Average concentration of CO is 30 mg/m³ for one hour and 10 mg/m³ for eight hours.

8.6 SULPHUR DIOXIDE (SO_2)

Sulphur dioxide (SO_2) is a colourless gas with a characteristic sharp pungent smell mainly produced by the burning of fossil fuels containing sulphur.

8.6.1 Sources of sulphur dioxide

(i) Burning of sulphur containing minerals and fossil fuels is the main source of SO_2.
(ii) Chemical plants release huge amount of SO_2 to the environment.
(iii) Exhaust of motor vehicles is another source of SO_2.
(iv) Power plant exhaust is another source of SO_2.

8.6.2 Adverse effects of sulphur dioxide

(i) SO_2 damages the lungs and eyes, causes coughing, wheezing and breathlessness, leading to bronchitis.
(ii) It is one of the main reactants to produce reducing smog.
(iii) Sufficient concentration of SO_2 brings about serious respiratory problems.
(iv) Marbles and building materials are affected in presence of sulphur dioxide.
(v) Presence of sulphur dioxide discolours clothes, leather and paper.
(iv) SO_2 undergoes photochemical reactions in the atmosphere to form H_2SO_4, which is ultimately precipitated on the Earth in the form of acid rain.

8.6.3 Permissible concentration of sulphur dioxide in air

The average permissible concentration of SO_2 is 350 μg/m³ for one hour and 120 μg/m³ for 24 hours.

8.7 OZONE (O_3)

Ozone is a colourless, highly reactive gas with a distinctive odour. It is formed naturally in the upper atmosphere (the stratosphere) and protects the Earth from ultraviolet (UV) radiation from the Sun. However, at ground level, high level of ozone is produced by reactions involving other air pollutants. Ozone is a **secondary air pollutant** formed by oxides of nitrogen (NO_2), O_2 and hydrocarbon reactions occurring in presence of sunlight.

8.7.1 Sources of ozone

(i) Emission of motor vehicles, combustion of fossil fuel.
(ii) Industrial and domestic use of solvents and coatings are the major source of ozone.
(iii) The nuclear waste is another important source of ozone.

8.7.2 Adverse effects of ozone

(i) Tropospheric ozone causes eye irritation, nose and throat irritation and breathing difficulties, leading to cardiovascular and respiratory diseases.
(ii) Exposure to 50 *ppm* ozone for several hours will lead to death due to accumulation of fluid in the lungs, leading to severe damage to lung capillaries. Younger people are susceptible to be affected more than older ones.
(iii) Ozone is responsible for malfunctioning of heart.
(iv) It damages trees and reduces growth rate.
(v) It reduces crop-yield, such as corn, wheat etc.

8.7.3 Permissible concentration of ozone in air

The average permissible concentration of ozone in air is less than 150 μg/m³ for one hour and less than 100 μg/m³ for eight hours.

8.7.4 Beneficial effects of ozone

Ozone in stratosphere region has the following beneficial effects on all living beings

(i) Ozone layer acts as a protective shield for living organisms on Earth from various injurious effects of UV rays (220–230 nm), which is proven to be responsible for sunburn and skin cancer.

(ii) In stratosphere ozone layer supplies necessary heat through absorption of UV rays to develop a positive lapse rate in the stratosphere due to which stratospheric region is quite stable and pollutants, if enters into the stratosphere, is entrapped locally instead of diffusion.

In stratosphere ozone to oxygen ratio is much higher than that in troposphere as shown below

$$\text{Troposphere: } \frac{O_2}{O_3} = 10^{25}, \qquad \text{Stratosphere: } \frac{O_2}{O_3} = 3 \times 10^5$$

8.8 HYDROCARBONS

Hydrocarbons are a class of reactive organic gases formed solely from hydrogen and carbon through incomplete burning of any organic matter such as oil, wood or rubber. Some examples of hydrocarbons are oil, petrol and solvents.

8.8.1 Sources of hydrocarbons

(i) Exhaust of combustion engine, oil refineries and power plants are the primary sources of hydrocarbons.

(ii) Evaporation from petroleum fuels (e.g. at service stations), solvents, dry cleaning solutions and paint are other sources of hydrocarbons.

(iii) Incomplete combustion in IC engine is another source of hydrocarbons.

(iv) Coal burning, forest fire etc.

(v) Exhausts of chemical and biochemical plants.

(vi) Anaerobic decomposition of organics in water, sediments and soil.

(vii) Human activities produces large quantities of hydrocarbons in the atmosphere.

(viii) Burning of papers, plastics, rubbers etc.

8.8.2 Adverse effects of hydrocarbons

(i) Hydrocarbons undergo photochemical reactions in presence of sunlight to form noxious pollutants.

$$CH_4 + O_2 \longrightarrow CH_3^*$$
$$CH_4 + OH^* \longrightarrow CH_3 + H_2O$$
$$CH_3\text{—}CH\text{=}CH_2 + OH^* \longrightarrow CH_3 - \dot{C}H\text{—}CH_2OH$$
$$CH_3\text{—}\dot{C}H - CH_2OH \longrightarrow CH_3CHO + CH_3^*$$

CH_3CHO is one of the important constituents of photochemical smog.

(ii) Methane (CH_4) reacts with mercury (Hg) to form CH_3Hg^+, which has detrimental effect on central nervous system and causes segregation and breakage of chromosomes, thereby inhibiting cell division. Only 5 *ppm* CH_3Hg^+ is enough to create such dreadful poisoning effects.

(iii) Hydrocarbons catalyse the oxidation of SO_2 to H_2SO_4 which is the main constituent of acid rain.

$$SO_2 + \frac{1}{2}O_2 + H_2O \xrightarrow{C_nH_m} H_2SO_4$$

(iv) High level of hydrocarbons in the atmosphere can reduce the amount of available oxygen.

8.8.3 Permissible concentration of hydrocarbons in air

(i) *Benzene* 3.6 μg/m³ for an annual average

(ii) 1,3-*Butadiene* 2.4 μg/m³ for an annual average (for 2002)

(iii) *Benzo-(α)-pyrene* 0.0003 μg/m³ for an annual average (for 2002)

8.9 SMOG

8.9.1 Photochemical smog

It is an oxidising smog, enriched with high level of oxidants, organics and appears as brown hazy fumes.

8.9.1.1 Constituents

The main constituents of photochemical smog are:

Nitrogen dioxide, aldehydes, free radicals, peroxybenzoyl/peroxyacyl nitrate (PAN) and reactive hydrocarbons, containing C=C groups.

8.9.1.2 Reactions

Hydrocarbons react with oxygen to produce hydrocarbon radicals, which undergo chain reactions as shown below

Reactive hydrocarbons

$$RCH_3 + O_2 \longrightarrow RCH_2^*$$

$$RCH_2^* + O_2 \longrightarrow RCH_2O_2^*$$

$$RCH_2O_2^* + NO \longrightarrow RCH_2O + NO_2$$

$$RCH_2O + O_2 \longrightarrow RCHO + HOO^*$$

$$HOO^* + NO \longrightarrow OH^* + NO_2$$

$$RCH_3 + OH^* \longrightarrow RCH_2^* + H_2O$$

Aldehydes react with OH^* in presence of NO_2 to form PAN as shown below

$$RCHO + OH^* \longrightarrow R\text{—}\overset{*}{C}\text{=O (acyl radical)} + H_2O$$

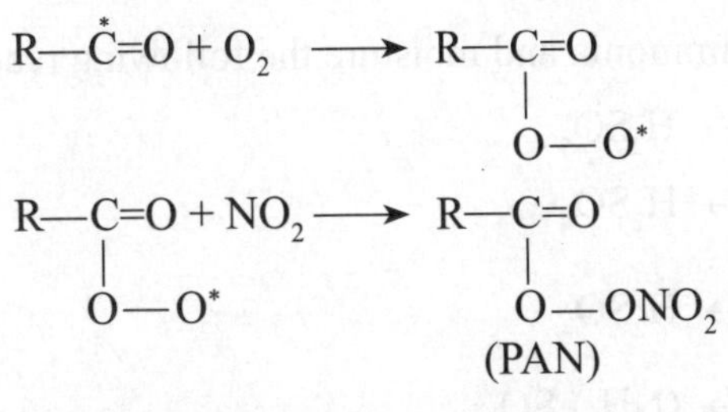

Peroxybenzoyl nitrate (PBzN) ≡ 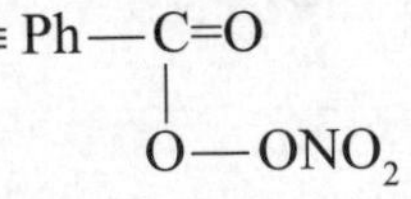

8.9.1.3 Conditions favourable for formation of photochemical smog

Usually high temperature around (35°C) is required to execute the above photochemical reactions. So midday summer is the ideal condition to form photochemical smog. Presence of SPM accelerates the formation of photochemical smog. This type of smog was first observed at Los Angeles, USA.

8.9.1.4 Adverse effects

(i) It causes irritation to eyes, damages lung capillaries.
(ii) Rubber articles get oxidised and cracked in presence of photochemical smog.
(iii) It also damages leaves, reduces plant life and crop yield.
(iv) It causes respiratory trouble.

8.9.2 Reducing smog

It is basically a mixture of sulphur dioxide and smoke. It is also known as London smog after the incident occurred in London in the 1950s. The main source is the burning of coal in power plants and burning of other fossil fuels. Illness, heart disease and respiratory problems are some of the major adverse effects of reducing smog. It also reduces visibility.

8.9.2.1 Constituents

The main constituents of reducing smog are $(NH_4)_2SO_4$, H_2SO_4 and moisture.

8.9.2.2 Reactions

The components, responsible for forming reducing smog, are sulphur oxides, suspended particulate matter, moisture and ammonia. Hydrocarbons, present in the atmosphere, act as catalyst

$$S + O_2 \longrightarrow SO_2$$

$$SO_2 + \frac{1}{2}O_2 \longrightarrow SO_3$$

In presence of hydrocarbon, ammonia and moisture the following reactions take place

$$SO_2 + H_2O \longrightarrow H_2SO_3$$

$$SO_3 + H_2O \longrightarrow H_2SO_4$$

$$H_2SO_3 + \frac{1}{2}O_2 \longrightarrow H_2SO_4$$

$$H_2SO_4 + NH_3 \longrightarrow (NH_4)_2SO_4$$

8.9.2.3 Adverse effects

(i) It damages the lungs and respiratory tracts, causing severe respiratory trouble and bronchitis.
(ii) It causes throat irritation.

8.9.2.4 Conditions favourable for formation of sulphurous smog

(i) Sufficient humidity.
(ii) Temperature inversion, i.e., temperature near the Earth surface must be low enough compared to that of air at higher altitude.
(iii) During early morning hours in a foggy winter season, ground air is sufficiently cooled and heavy so that vertical mixing cannot take place. Thus pollutants like SO_x are trapped near the ground and undergoes the above chemical reactions to form reducing smog.

8.10 ACID RAIN

Exhausts of power plants and chemical plants as well as burning of fuel generates huge smoke, consisting of harmful gases like sulphur dioxide, nitrogen dioxide, chlorine etc. and soot particles. These gases react with water droplets of clouds to form weak acid solution, which precipitates on the Earth as rain, known as acid rain. The pH of acid rain lies within 2 to 4 whereas pH of normal rain lies within 5 to 6. The following photochemical reactions are believed to occur in the atmosphere in presence of hydrocarbons.

$$2SO_2 + 2H_2O + O_2 \longrightarrow 2H_2SO_4$$

$$4NO_2 + 2H_2O + O_2 \longrightarrow 4HNO_3$$

$$Cl_2 + H_2O \longrightarrow HCl + HOCl$$

8.10.1 Adverse effects of acid rain

(i) It adversely affects the soil and plant growth.

(ii) Water of aquatic reservoir becomes acidic and hence aquatic life becomes endangered.

(iii) Acid rain dissolves and removes nutrients and minerals in the soil.

(iv) Acid rain is a good carrier of harmful chemicals and metals, which are responsible for fast corrosion of soil.

(v) Buildings and materials, made of limestone or sandstone, are worst affected by the acid rain.

(vi) Cable lines and pipe lines are also affected by the acid rain.

8.10.2 Precautionary measurement to prevent acid rain

(i) Exhausts of power plants, factories and vehicles should be treated properly before releasing to the atmosphere.

(ii) Burning of fuels should be restricted.

9

Air Pollution Control

9.1 CRITERIA POLLUTANTS

National Ambient Air Quality Standards (NAAQS) has identified *six* criteria pollutants. The pollutants and their permissible limits in the environment are listed in the following table. One of the criteria pollutants is SPM, which comprises of PM_{10} and $PM_{2.5}$.

Table 9.1 Criteria pollutants and their permissible limits.

Pollutants	*Permissible limit*
Carbon monoxide	9 *ppm* for 8 hours
Nitrogen oxides	0.05 *ppm* annually
Sulphur dioxide	0.03 *ppm* annually
PM_{10}	150 $\mu g/m^3$ for 24 hours
$PM_{2.5}$	35 $\mu g/m^3$ for 24 hours
Ozone	0.12 *ppm* for 1 hour
Lead	0.15 $\mu g/m^3$ for 3 months

Factory exhaust and automobile exhaust are the main sources of these pollutants. Thus these exhausts need to be treated before releasing to the atmosphere.

9.2 TREATMENT OF FACTORY EXHAUST

Factory exhaust is sent through a series of equipment to remove harmful particles and gases and ultimately released to the atmosphere. The different steps are discussed below.

9.2.1 Cyclone separator

The exhaust is first sent to a cyclone separator, where it is subjected to a centrifugal field. Cyclone separator is a centrifugal separator where the particles are swung as a result of their mass by the centrifugal force. Large particles are accelerated towards the wall of separator and hurl on it. Ultimately the particles slide down to the bottom and deposited at the dust collector, attached at the bottom. Clean gas is taken out from the top. Large particles, having diameter ≥10 μm can be separated successfully by using this technique. Dirty air enters into the cyclone separator at an average speed of 16 m/s. The diameter of particles, separated and settled in the cyclone separator, can be computed by using the following formula

$$d_{50} = \sqrt{9\mu_i b / 2\pi N \Gamma_i v_i}$$

Where,

d_{50} = Particle size (m), separated from the gas stream.
b = Inlet width of the cyclone separator (m).
N = Number of turn cycles in the cyclone separator. Usually N = 5 is applied.
v_i = Inlet speed of the gas stream (m/s).
Γ_i = Specific mass of the inlet dust particle (kg/m³).
μ_i = Viscosity of the inlet gas (kg/(m-s)).

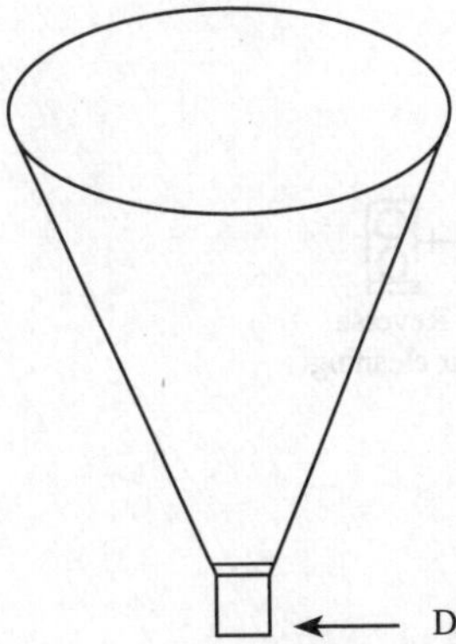

Fig. 9.1 Schematic diagram of cyclone separator.

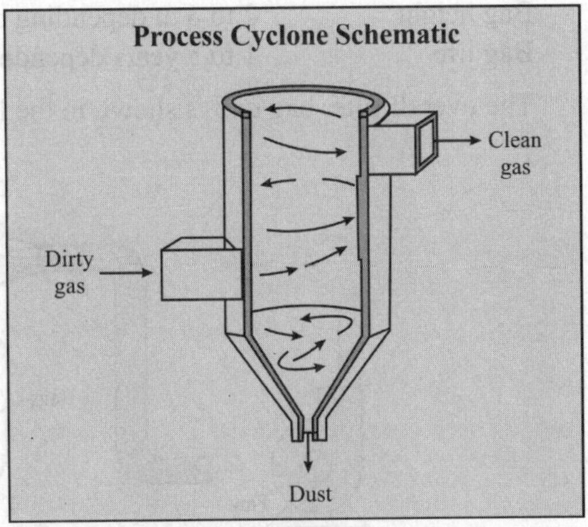

Fig. 9.2 Schematic diagram of separation process.

9.2.2 Filter bag unit

The dirty exhaust is then sent to a filter bag unit, consisting of several filter bags, suspended from the top. The dirty gas is forced to pass through these bags from the bottom and comparatively clean gas is taken out from the top. During passing through the filter bags, particles, having diameter ≥ 5 μm, are separated and collected at the dust collector, attached at the bottom. The schematic diagram of filter bag unit is shown in the Fig. 9.3.

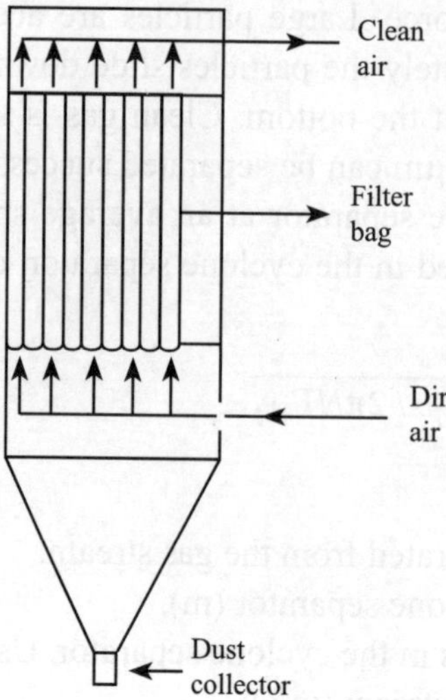

Fig. 9.3 Schematic diagram of filter bag unit.

Bag diameter range	0.1 to 0.35 m.
Bag height	2 to 6 m depending on type of products and production rate.
Bag life	1 to 5 years depending on the usage.

The overall filter bag unit is shown in the Fig. 9.4.

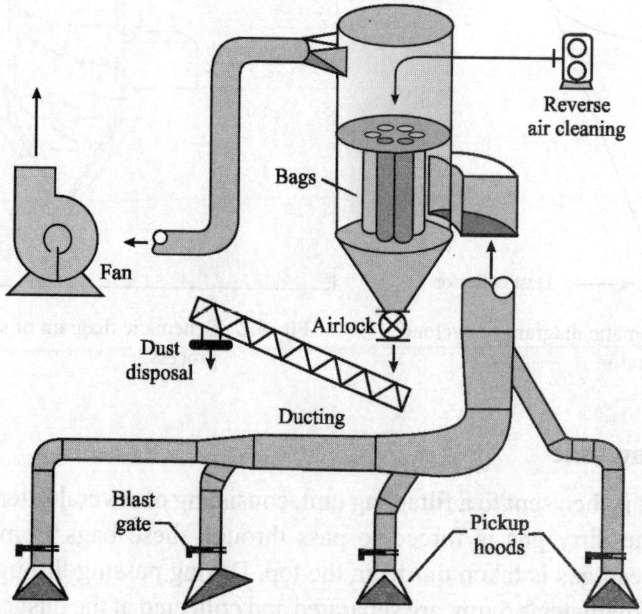

Fig. 9.4 Schematic diagram of overall filter bag unit.

Bags are usually made of different materials and operating temperature depends on filter bag materials. Characteristics of different fabrics are shown in the Table 9.2.

Table 9.2 Characteristics of different fabrics.

Fabric	*Maximum working temperature (°C)*	*Acid resistance*	*Basic resistance*
Cotton	80	Poor	Good
Wool	90	Good	Poor
Nylon	90	Poor	Good
Polypropylene	90	Excellent	Excellent
Dacron	135	Good	Fair
Teflon	200	Excellent	Excellent
Glass	280	Good	Good

9.2.2.1 Efficiency of a baghouse depends on the following factors

(i) Types of baghouse.
(ii) Filtering velocity applied.
(iii) Dust loading in the exhaust.
(iv) Filter bag material.
(v) Filter drag model, which helps increase the efficiency.

9.2.2.2 Advantages of filter bag unit

(i) High efficiency to remove small particles from exhaust.
(ii) Different types of modular design are available.
(iii) Low velocity is enough.
(iv) Low pressure drop is sufficient.

9.2.2.3 Disadvantages of filter bag unit

(i) Large area is required to set up the unit.
(ii) Frequent cleaning of filter bag unit is necessary to execute smooth running.
(iii) As the material of filter bags is temperature sensitive, operating temperature limitations is another disadvantage.

9.2.2.4 Applications

The filter bag unit is widely used in the following industries

(i) Power plants.
(ii) Fertilizer plants.
(iii) Steel mills and paper industries.
(iv) Food processing units.
(v) Cement and asbestos manufacturing units.

9.2.3 Liquid scrubbers

The exhaust is then sent to two successive liquid scrubbers, namely, alkaline scrubber and acid scrubber. While passing through the alkaline scrubber acidic oxides, e.g., carbon dioxide, sulphur dioxide, nitrogen dioxide etc. are absorbed and alkaline oxides, e.g., alumina, silica etc. are absorbed in the acid scrubber. The clean air is then sent to wet scrubber, containing $CaCO_3$ slurry. While passing through this scrubber remaining SO_2 gas reacts with $CaCO_3$ to form the crystal $CaSO_3 \cdot 3H_2O$, which is coming out of the scrubber along with $CaCO_3$ slurry. The clean air, coming from the wet scrubber, is sent to the next unit.

The schematic diagram of scrubber unit is shown in the Fig. 9.5. The Figs. 9.6, 9.7 and 9.8 illustrate the model of different scrubber units, used in industries. There is different type of scrubbers, e.g., Jet scrubber, Venturi scrubber etc.

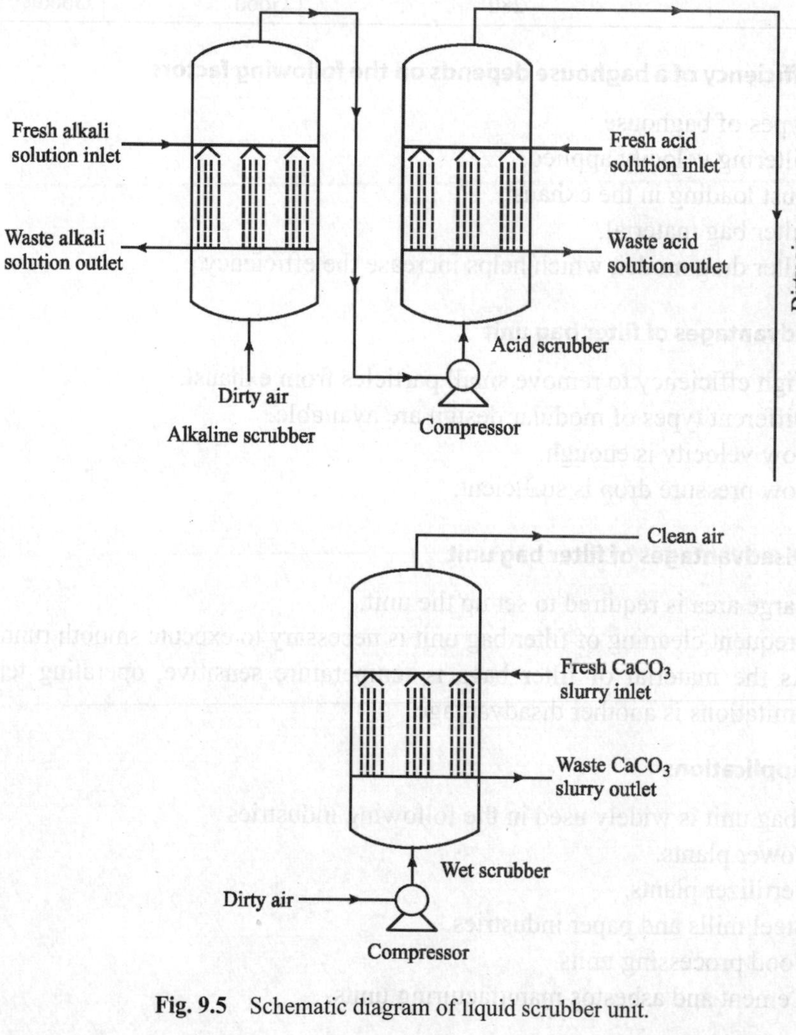

Fig. 9.5 Schematic diagram of liquid scrubber unit.

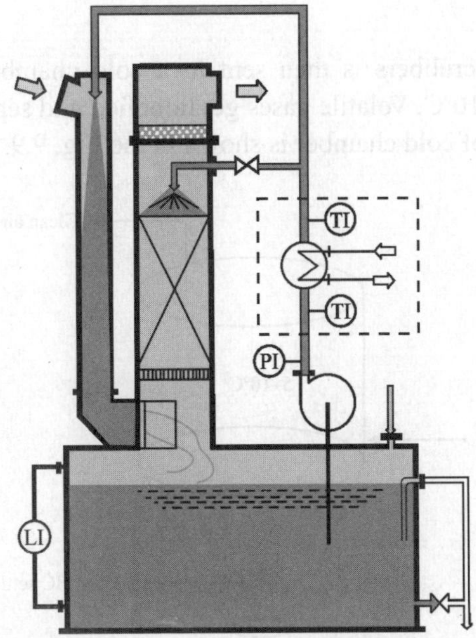

Fig. 9.6 Industrial liquid scrubber unit.

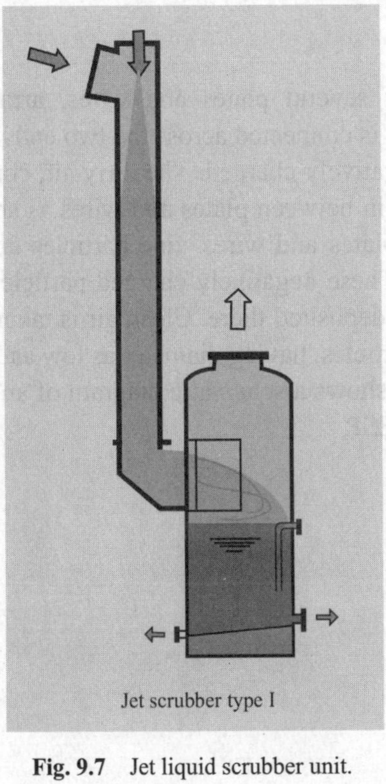

Fig. 9.7 Jet liquid scrubber unit.

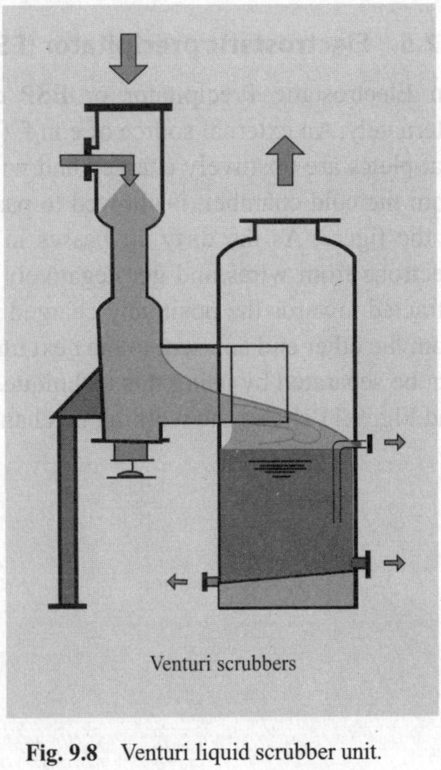

Fig. 9.8 Venturi liquid scrubber unit.

9.2.4 Cold chamber

The exhaust of liquid scrubbers is then sent to a cold chamber where temperature is maintained at around 5–10°C. Volatile gases get liquefied and separated from the exhaust. The schematic diagram of cold chamber is shown in the Fig. 9.9.

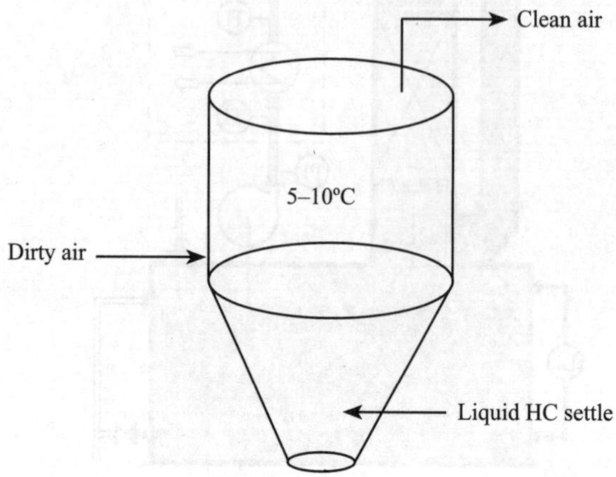

Fig. 9.9 Schematic diagram of cold chamber.

9.2.5 Electrostatic precipitator (ESP)

An Electrostatic Precipitator or ESP consists of several plates and wires, arranged alternately. An external source of e.m.f. (30–75 kV) is connected across the two ends such that plates are positively charged and wires are negatively charged. The dirty air, coming from the cold chamber, is allowed to pass through in between plates and wires as shown in the figure. As the dirty air passes in between plates and wires, fine particles collect electrons from wires and get negatively charged. These negatively charged particles are attracted towards the positively charged plates and deposited there. Clean air is taken out from the other end and sent to the next unit. Fine particles, having diameter as low as 1 μm can be separated by using this technique. Fig. 9.10 shows a schematic diagram of an ESP and Fig. 9.11 shows the working mechanism of an ESP.

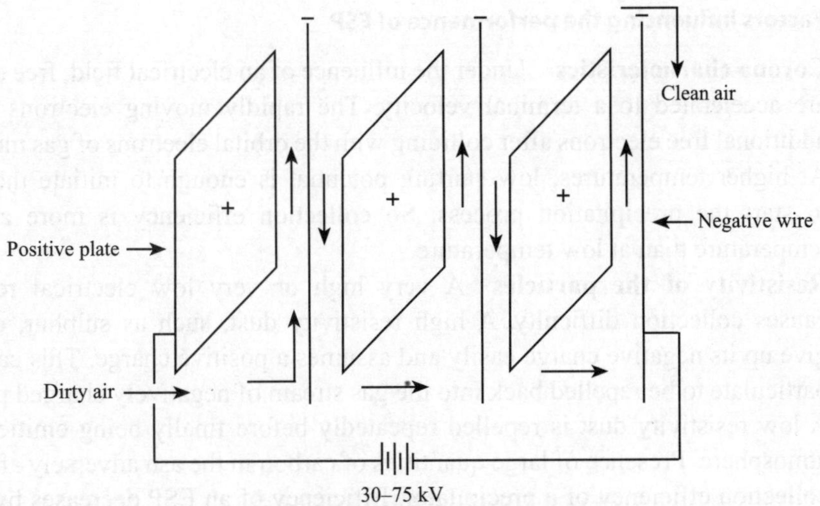

Fig. 9.10 Schematic diagram of an ESP.

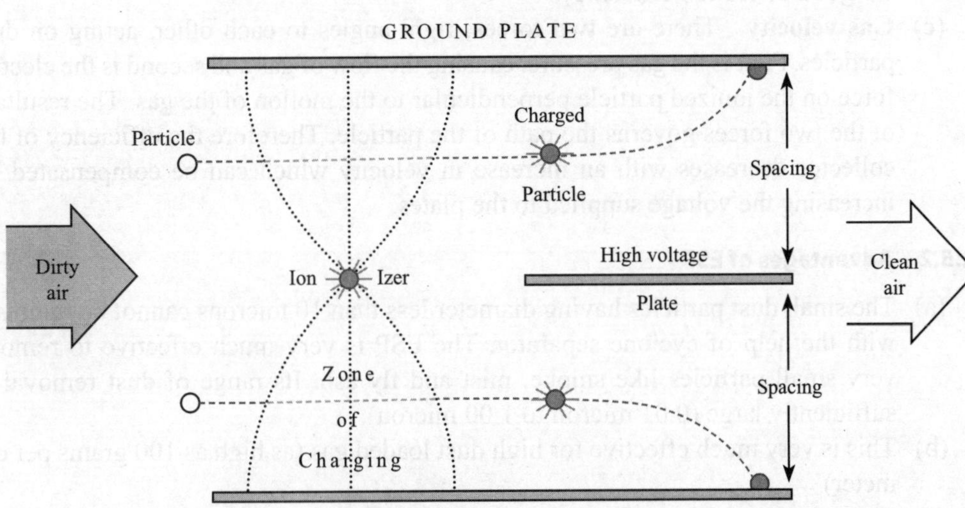

Fig. 9.11 Mechanism of separation of particles in an ESP.

Efficiency (η) of ESP is given by $\eta = 1 - e^{-A.v/Q}$

where, η = the efficiency of ESP.

A = the cross-sectional area of the plate (m^2).

v = the velocity of the gas stream (m/s).

$\dot{Q}$ = the volumetric flow rate of the gas stream (m^3/s).

v is constant for a given potential difference and $\dot{Q}$ is also constant for a given exhaust system. The value of A can be changed and η increases with increase in value of A. Thus efficiency of an ESP increases with increase in plate area.

9.2.5.1 Factors influencing the performance of ESP

(a) **Corona characteristics** Under the influence of an electrical field, free electrons are accelerated to a terminal velocity. The rapidly moving electrons produce additional free electrons after colliding with the orbital electrons of gas molecules. At higher temperatures, low starting potential is enough to initiate the corona to start the precipitation process. So collection efficiency is more at higher temperature than at low temperature.

(b) **Resistivity of the particles** A very high or very low electrical resistivity causes collection difficulty. A high resistivity dust, such as sulphur, does not give up its negative charge easily and assumes a positive charge. This causes the particulate to be repelled back into the gas stream of negatively charged particles. A low resistivity dust is repelled repeatedly before finally being emitted to the atmosphere. Presence of large quantities of carbon in the ash adversely affects the collection efficiency of a precipitator. Efficiency of an ESP decreases by 1% for every 1% of carbon in the gas over 15%. Therefore, medium resistivity is required for good collection efficiency.

(c) **Gas velocity** There are two forces, right angles to each other, acting on dust particles. First is the gas pressure, causing the flow of gas and second is the electric force on the ionized particle perpendicular to the motion of the gas. The resultant of the two forces governs the path of the particle. Therefore the efficiency of the collector decreases with an increase in velocity which can be compensated by increasing the voltage supplied to the plates.

9.2.5.2 Advantages of ESP

(a) The small dust particles having diameter less than 10 microns cannot be removed with the help of cyclone separator. The ESP is very much effective to remove very small particles like smoke, mist and fly ash. Its range of dust removal is sufficiently large (0.01 micron to 1.00 micron).

(b) This is very much effective for high dust loaded gas (as high as 100 grams per cu. meter)

(c) The operation is easy.

(d) The dust is collected in dry form and can be removed either in dry form or in wet form.

(e) The ESP can be designed to have more than 99% collection efficiency.

(f) Silicon control rectifiers and other modern control devices allow an electrostatic precipitator to operate automatically.

(g) Low maintenance cost.

9.2.5.3 Disadvantages of ESP

(a) Very high voltage is required to execute an ESP. So, modern electrical equipment is necessary to convert low voltage (400 V) AC to high voltage (60000 V) DC.

(b) High capital cost to provide high voltage.

(c) Large space is required to install an ESP.
(d) The efficiency of the collector depends on the gas velocity. So gas velocity is to be maintained properly. The dust carried with the gases increases with an increase in gas velocity.
(e) As high potential is used, it is necessary to protect the entire collector from sparking by providing a fine mesh before the ionizing chamber. This is necessary because even a smallest piece of paper might cause sparking when it would be carried across adjacent plates or wires.
(f) No SO_2 control is possible with an ESP.
(g) An ESP is sensitive to its design parameters.

9.2.6 Thermal precipitator

The thermal precipitator is a closed chamber, consisting of several zones, separated by insulated layer. Temperature at the bottom zone is the lowest. As we move upward, temperature increases and the temperature is the highest at the top zone. The dirty air, coming from the ESP is passed through the precipitator from bottom. As the dirty air passes from lower zone to higher zone, temperature increases and hence density of dirty air decreases. Thus fine particles are slowly separated out. Very fine particles, having diameter as low as 0.1 μm can be separated by using this technique. Clean air is taken out from the top and sent to the absorption column. The schematic diagram of a thermal precipitator is shown in the Fig. 9.12.

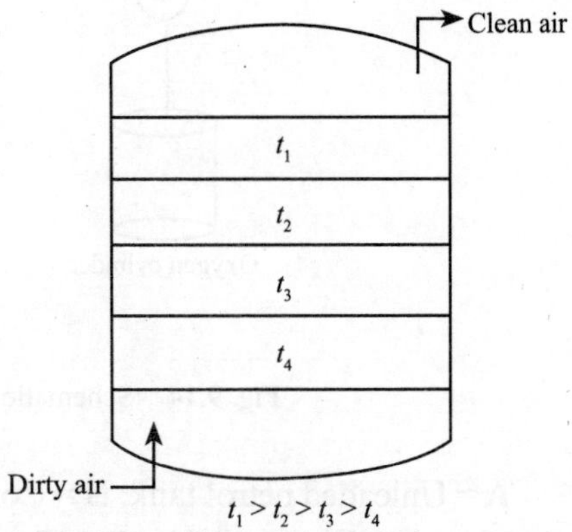

Fig. 9.12 Schematic diagram of a thermal precipitator.

9.2.7 Absorption column

The gas stream, coming from thermal precipitator, is passed through an absorption column from the bottom. The column is filled with alumina (Al_2O_3) and silica gel mixture, which absorbs undesired hydrocarbons from the gas stream. Finally the gas is released to the atmosphere. The absorption column is shown in the Fig. 9.13.

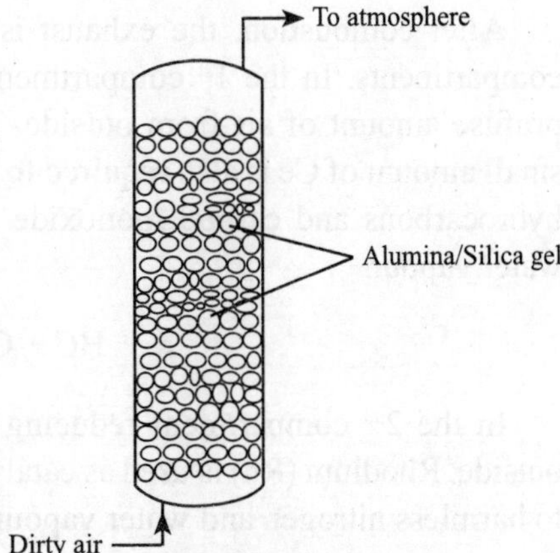

Fig. 9.13 Schematic diagram of an absorption column.

9.3 TREATMENT OF AUTOMOBILE EXHAUST

Automobile exhaust is a big source of air pollution and hence the exhaust should be treated properly before releasing to the atmosphere. The following figure (Fig. 9.14) illustrates the in-situ treatment process of automobile exhaust.

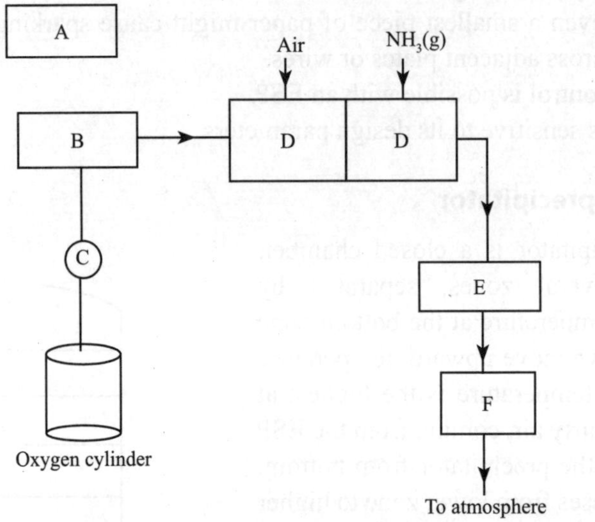

Fig. 9.14 Schematic diagram of automobile exhaust treatment.

A = Unleaded petrol tank, B = Combustion chamber, C = Oxygen sensor, D = Catalytic chamber, E = Charcoal chamber, F = Silencer.

In automobiles unleaded petrol is recommended in order to avoid lead pollution. The unleaded petrol is first sent to a combustion chamber, B, where air fuel ratio is maintained through an oxygen sensor, C, which is attached to an oxygen cylinder.

After combustion, the exhaust is sent to a catalytic chamber, D, consisting of two compartments. In the 1st compartment oxidising atmosphere is maintained by supplying profuse amount of air from outside. Finely divided Pt/Pd mixture is used as catalyst. A small amount of Ce is also required to promote the oxidation process. Under this condition, hydrocarbons and carbon monoxide are oxidised to form harmless carbon dioxide and water vapour.

$$HC + CO \longrightarrow CO_2 + H_2O$$

In the 2nd compartment reducing atmosphere is created by injecting ammonia from outside. Rhodium (Rh) is used as catalyst. Under this condition, nitrogen oxides are reduced to harmless nitrogen and water vapour.

$$NO_x + [H] \longrightarrow N_2 + H_2O$$

Now-a-days Three-Way-Catalyst (TWC), where both the oxidation and reduction processes take place simultaneously, is used. TWC chamber is shown in the Fig. 9.15.

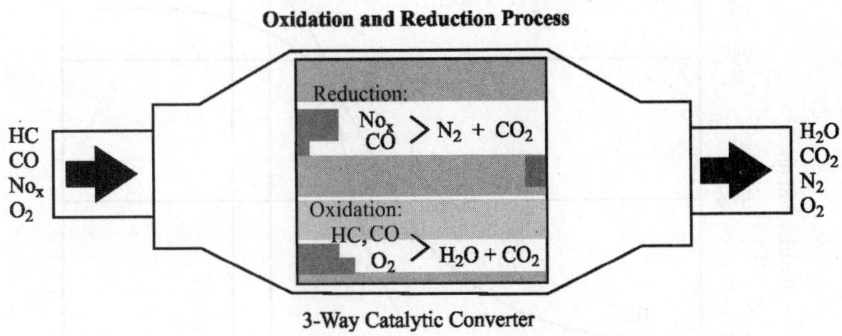

Fig. 9.15 Schematic diagram of a catalytic converter.

The exhaust is then sent to a charcoal chamber, where remaining unburnt hydrocarbons are absorbed. Finally the exhaust is released to the atmosphere through a silencer to minimize noise pollution.

9.3.1 Efficiency of a catalytic converter

Efficiency of a cataylic converter depends on temperature and feed gas composition.

(i) Minimum temperature required to start oxidation process is nearly 280°C but efficient purification occurs when temperature reaches around 400°C.

(ii) If air content in air/fuel mixture is higher than that required in stoichiometric equation, the exhaust contains sufficient amount of oxygen with little amount of carbon monoxide. So good oxidising environment is developed within the catalytic converter and the oxidising catalyst mixture, containing Pt, Pd and Ce, becomes very much efficient and the excess oxygen is stored.

(iii) If air content in air/fuel mixture is lower than that required in stoichiometric equation, the exhaust contains sufficient amount of carbon monoxide with little amount of oxygen. Thus reducing environment is developed within the catalytic converter and the reducing catalyst, containing Rh, becomes very much efficient and the stored oxygen is released.

In order to have efficient catalyst performance air/fuel ratio is to be maintained around 14.7:1. Fig. 9.16 illustrates how purification rate depends on air/fuel ratio.

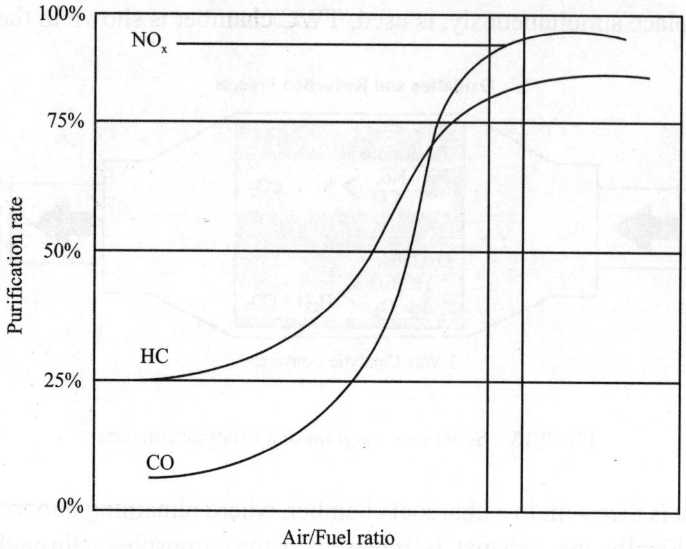

Fig. 9.16 Air/Fuel ratio in a catalytic converter.

Oxygen sensor plays a vital role to maintain oxygen content in the air/fuel mixture. The sensor may be affected due to the following reasons

(i) Silicon contamination. Si burns in the combustion chamber to form SiO_2 glaze, which affects the efficiency of oxygen sensor.

(ii) Lead contamination.

(iii) Carbon contamination.

10

Water Pollution

10.1 HYDROLOGIC CYCLE

(a) *Evaporation* During day time huge amount of water from natural water resources like ponds, rivers, oceans etc. get evaporated in natural process.

(b) *Transpiration* Plants extract water from soil through their roots and this water is then transported to the leaves from which it is finally evaporated. The phenomenon is called transpiration.

(c) *Condensation* In winter season water vapour in atmosphere is condensed into droplets of liquid water in the form of fog and dew. The phenomenon is called condensation.

(d) *Precipitation* Heavy waterfall occurs on the surface of the Earth in the form of rain, hail, snow, sleet etc. The phenomenon is called precipitation.

(e) *Infiltration* Movement of water from one aquifer to other confined water source is called infiltration.

(f) *Percolation* In many cases water from surface water sources moves to the surface through pores of soil. The phenomenon is called percolation.

(g) *Run-off* Due to melting of ice caps, downflow of water is observed from hill area to the sea level. The phenomenon is called run-off.

10.2 CAUSES OF WATER POLLUTION

(a) Dumping of wastes and chemicals in the urban and rural watersheds.

(b) Excessive use of fertilizers and pesticides in the agricultural land. Farm run-off, landfill sites are some major contaminating sources of fertilizers and pesticides. Extra nutrients in water causes eutrophication in several water resources, like ponds, lakes, rivers etc. Excessive use of fertilizers causes nitrate contamination of groundwater.

(c) Continuous discharge of effluents, containing hazardous chemicals from industries, to the nearby water reservoir. Several pollutants, like sulphur dioxide, nitrogen dioxide etc., released from chemical and power plants, are carried by wind and ultimately mixed with the surface water resources. These pollutants form acids in water, thereby, increasing the acidity.

(d) Municipal waste sewage is another source of water pollution in the developing countries. As the municipal waste possesses significant amount of organics, bacteria consumes substantial amount of dissolved oxygen for biodegradation of organics. Furthermore, the waste carries substantial pathogens, responsible for water-borne diseases.

(e) Many of the 100000 synthetic compounds in use today are found in the aquatic environment and accumulate in the food chain. Persistent organic pollutants (POPs) represent the most harmful element for the ecosystem and for human health. Examples are industrial chemicals and agricultural pesticides. Fishes consume these chemicals, which cause serious damage to human health.

(f) Leakage in pipelines and tanks.

(g) Sudden spill over of oils and chemicals into the water reservoir.

(h) Lack of social responsibilities.

10.3 ENVIRONMENTAL DAMAGES DUE TO WATER POLLUTION

(a) Contaminated water threatens the health of millions of people through spreading of water-born diseases.

(b) Erosion of soil and degradation of soil quality, deforestation, resulting in huge loss in national productivity.

(c) Pollution in coastal water, rivers, lakes, ponds etc. immensely damages the aquatic life leading to a remarkable reduction in fish population.

(d) Industrial effluent, enriched with As, Cd, Hg, Pb etc., is poisonous to aquatic plants and animals. These dreadful elements spread to human body through consumption of contaminated fishes etc.

(e) Fluoride containing pollutants are responsible for gastrointestinal problems, dental problems, respiratory problems etc.

(f) Oil containing pollutants have poisonous effect on birds and fishes.

Steps to be taken to prevent water pollution

(a) Use of fertilizers and pesticides in agricultural sectors should be restricted.

(b) Industrial and municipal effluents should be discharged to the nearby reservoirs only after proper treatment.

(c) Sufficient supply of water to all sectors like industrial sectors, domestic areas and agricultural sectors.

(d) Minimum water quality standard should be maintained in drinking water as prescribed by WHO.

(e) Effort should be made to discover new water resources.

10.4 DECOMPOSITION OF ORGANIC WASTES

Aerobic decomposition

Microorganisms have a natural tendency to degrade or decompose organic wastes into simpler organic and inorganic substances. Some microorganisms decompose the organic wastes in presence of oxygen atmosphere. The process is known as aerobic decomposition and the corresponding bacteria is called aerobic bacteria. The products of decomposition are usually CO_2, H_2O, NO_3^-, PO_4^{3-}, SO_4^{2-} etc., which are non-objectionable.

Anaerobic decomposition

Some microorganisms decompose the organic wastes in absence of oxygen atmosphere. The process is known as anaerobic decomposition. The products of decomposition are usually PH_3, H_2S, CH_4 etc., which are foul smelling and objectionable.

10.5 BIOCHEMICAL OXYGEN DEMAND (*BOD*)

The total amount of dissolved oxygen required by aerobic bacteria to decompose all the organic wastes of a given waste water sample aerobically is called biochemical oxygen demand or *BOD*. The unit of *BOD* is *mg/l*.

Significance For a given waste sample higher *BOD* value indicates larger amount of decomposable organic matter present in the waste and hence the waste is said to be more polluted. So, *BOD* value indicates the degree of pollution in a given waste sample. The water quality of a waste, having *BOD* value lying within 2, is *very good*, the waste, having *BOD* value lying within 8, is taken as *moderately clean waste*. When *BOD* value crosses 10, the water quality is *poor* and the waste is said *somewhat polluted*. When *BOD* value of a waste sample reads 100 or more, the waste is said to be *very much polluted*. So *BOD* value indicates the pollution index of the given water sample.

10.5.1 The BOD_5 test

The rate of oxygen consumed by microorganisms in a given test sample of water is measured at a fixed temperature (usually 20°C) and over a given time period in the dark. The minimum time period required for *BOD* test is five days, but other *BOD* tests are also used. There are several methods for the measurement of *BOD*.

10.5.2 Unseeded dilution method

In this method a given volume of sample is taken in an amber coloured stoppered bottle, marked 300 ml. The sample is diluted up to the mark by deionized water, containing no other bacteria. The initial dissolved oxygen (*DO*) content is measured. It is denoted by DO_i.

The bottle is kept in the dark for five days. The following precautionary measures are to be taken

(a) The bottle must be stoppered properly in order to avoid oxygen diffusion since it is required that microorganisms consume dissolved oxygen only to eat up organics of the sample.

(b) The bottle must be coloured and kept in dark in order to avoid any photosynthesis.

After at least five days the bottle is taken out and the final *DO* content is measured. It is denoted by DO_f. The BOD_5 is given by

$$BOD_5 = \frac{DO_i - DO_f}{P} \tag{10.1}$$

where, *P* is called *dilution factor* and is given by: $P = \frac{V}{300}$ and *V* is the volume of the sample taken.

10.5.3 Seeded dilution method

In this method seeded water, i.e., deionized water contaminated with aerobic bacteria, is used. Two bottles are taken. In the 1st bottle a given volume of sample (V_w ml) diluted up to the mark with deionized water, containing aerobic bacteria, is taken. The 2nd bottle is filled up to the mark by only deionized water, containing aerobic bacteria. Both the bottles are kept in dark for at least five days.

$DO_i(1)$ = Initial *DO* content of the 1st bottle. It is a measurable quantity with unit *mg/l*.
$DO_i(2)$ = Initial *DO* content of the 2nd bottle. It is a measurable quantity with unit *mg/l*.

After five days two bottles are taken out and *DO* content of the bottles are measured.

$DO_f(1)$ = Final *DO* content of the 1st bottle. It is a measurable quantity with unit *mg/l*.
$DO_f(2)$ = Final *DO* content of the 2nd bottle. It is a measurable quantity with unit *mg/l*.

So, *BOD* of the mixture = $BOD_m = DO_i(1) - DO_f(1)$. It is known.
BOD of the deionized water = $BOD_d = DO_i(2) - DO_f(2)$. It is known.

V_w = Volume of the waste sample taken in litre.
V_m = Volume of the waste sample plus seeded deionized water in litre.
V_d = Volume of the deionized water in litre.

Thus, $$V_m = V_w + V_d \quad \text{and} \quad \frac{V_w}{V_m} = P, \quad \frac{V_d}{V_m} = 1 - P$$

Where, *P* is called dilution factor.

Again, total mass of oxygen remains constant. So, we can write that

$$BOD_m \times V_m = BOD_w \times V_w + BOD_d \times V_d$$

Dividing both sides by V_m, we have,

$$BOD_m = BOD_w \times P + BOD_d \times (1 - P)$$

Or, $$BOD_w = \frac{BOD_m - BOD_d \times (1-P)}{P} \quad (10.2)$$

10.5.4 Manometric method

In this method the amount of oxygen consumed by aerobic bacteria only due to carbonaceous oxidation is determined. Nitrogenous oxidation is inhibited by using inhibitors. So only carbonaceous oxygen demand, i.e., *CBOD* can be determined fairly accurately.

Procedure A given volume of water sample is kept in a coloured and stoppered bottle with a pressure sensor. A strong carbon dioxide absorbent (typically *LiOH*) is added in the bottle such that the absorbent lies above the sample level. The sample is stored in conditions identical to the dilution method. Aerobes consume dissolved oxygen and carbon dioxide is released. As carbon dioxide gas is absorbed, the pressure decreases. The electronic measurement of this drop of pressure gives the consumed quantity of oxygen. Allylthiourea is generally added with the sample in order to avoid oxidation of nitrogeneous matter.

Advantages (i) No dilution of sample is required. (ii) Seeding is not required. (iii) No blank experiment for seeded sample is required. (iv) Direct reading of *BOD* is obtained.

10.5.5 The *BOD* reaction rate

It is believed that biochemical oxidation reaction follows the 1st order kinetics.

BOD_t = The concentration of dissolved oxygen already consumed by aerobes to decompose the waste in a given time period *t*. Its unit is *mg/l*.

L_t = The concentration of dissolved oxygen remains unconsumed by aerobes to decompose the waste after the time interval *t*. In other words, L_t is called remaining oxygen demand. Its unit is *mg/l*.

L_0 = The ultimate oxygen demand by aerobes to decompose all the carbonaceous waste aerobically. Its unit is *mg/l*. So, $L_0 = BOD_t + L_t$.

Thus according to 1st order kinetics, the *BOD* reaction rate is given by

$$-\frac{dL_t}{dt} \propto L_t$$

Or, $$-\frac{dL_t}{dt} = k_d L_t,$$

where, k_d is the biodegradation reaction rate constant with unit day^{-1}.

Boundary conditions

when, $t = 0, L_t = L_0$ and when, $t = t, L_t = L_t$

On integration of the above equation with boundary conditions, we get

$L_t = L_0 e^{-k_d t}$. Thus BOD_t is given by $BOD_t = L_0 - L_0 e^{-k_d t}$

$$BOD_t = L_0 - L_0 e^{-k_d t} \text{ and } L_t = L_0 e^{-k_d t} \quad (10.3)$$

The Fig. 10.1 illustrates the change in BOD_t and L_t with time.

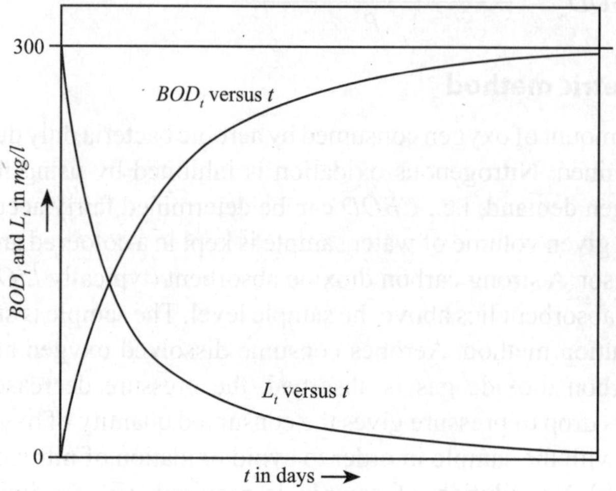

Fig. 10.1 Variation of BOD_t and L_t with time.

The deoxygenation rate constant, k_d depends on temperature. The standard value of k_d is 0.22 day^{-1} at 20°C and it increases with temperature according to the following equation

$$k_\theta = k_{20}(\alpha)^{\theta-20}$$

Where, k_θ is the deoxygenation rate constant at any temperature θ and α is a constant and its value is 1.047.

10.5.6 Disadvantages in *BOD* determination

(a) *BOD* results sometimes appear erroneous due to the presence of some undesirable substances.
(b) Some organic materials such as cellulose, phenols, benzene and tannic acid do not respond biodegradation process.
(c) Some organic materials, such as pesticides, industrial chemicals are non-biodegradable because they are toxic to microorganisms.
(d) Determination of *BOD* requires long time, at least five days.
(e) The *BOD* reaction curve shows that biodegradation takes infinite time to complete. Thus *BOD* result is always inaccurate.

10.6 CHEMICAL OXYGEN DEMAND (*COD*)

It refers to the total amount oxygen required to decompose all the organic waste of a given water sample by chemical means. The unit of *COD* is *mg/l*.

The following reagents are used as oxidising agents: (i) potassium dichromate, (ii) sulphuric acid, (iii) trace amount of silver sulphate, (iv) trace amount of mercuric sulphate, (v) Standard Mohr's salt solution, (vi) ferroin indicator.

10.6.1 Determination of *COD*

Standard method of determination of *COD* of a given waste sample is described in ISO 6060.

(a) Several oxidising agents may be used for the *COD* test, e.g., potassium permanganate, ceric sulphate, potassium iodate and potassium dichromate. However, potassium dichromate appears the most effective reagent to oxidise almost all organic compounds. That is why, potassium dichromate ($K_2Cr_2O_7$) with 50% sulphuric acid (H_2SO_4) is used to oxidise the wastes. The following reaction takes place during oxidation process

$$C_nH_aO_bN_c + dCr_2O_7^{2-} + (8d+c)H^+ \rightarrow nCO_2 + \frac{a+8d-3c}{2}H_2O + cNH_4^+ + 2dCr^{3+}$$

where, $d = \frac{2n}{3} + \frac{a}{6} - \frac{b}{3} - \frac{c}{2}$. Usually, 0.25 N $K_2Cr_2O_7$ solution, containing 1:1 H_2SO_4, is used for *COD* determination. However, for samples, having *COD* below 50 *mg/l*, a lower concentration of potassium dichromate solution is preferred.

(b) Trace amount of silver sulphate (Ag_2SO_4) is required to catalyse the oxidation of straight chain aliphatic compounds, aromatic hydrocarbons and pyridine.

(c) Trace amount of mercuric sulphate ($HgSO_4$) is used. Hg^{2+} ion forms a complex ($HgCl_4^{2-}$) with Cl^-, preventing the interference of the latter during *COD* determination. Otherwise, Cl^- is oxidised by potassium dichromate according to the following reaction

$$6Cl^- + Cr_2O_7^{2-} + 14H^+ \longrightarrow 3Cl_2 + 2Cr^{3+} + 7H_2O$$

(d) Excess dichromate is always present in the system after completion of oxidation process. This excess potassiumdichromate is estimated with the help of standard Mohr's salt solution [$FeSO_4.(NH_4)_2SO_4.6H_2O$] by using ferroin indicator, which changes colour from blue-green to reddish-brown colour at the end point. *COD* of the sample can be calculated from the following equation

$$COD = \frac{8000(b-s)N}{V} \, mg/l \tag{10.4}$$

Where,

b = Volume of Mohr's salt consumed during blank titration in ml.

s = Volume of Mohr's salt consumed by excess dichromate after oxidation in *ml.*

N = Normality of the Mohr's salt solution (*gm·eq./l*).

V = Volume of the waste water sample in ml.

Equivalent weight of oxygen is 8.

COD indicates the amount of oxygen required to oxidize an organic waste into carbon dioxide, ammonia and water. The following reaction is believed to occur

$$C_nH_aO_bN_c + \left(n + \frac{a}{4} - \frac{b}{2} - \frac{3c}{4}\right)O_2 \longrightarrow nCO_2 + \left(\frac{a}{2} - \frac{3c}{2}\right)H_2O + cNH_3$$

However, the above expression does not include the oxidation of ammonia to nitrate, which is commonly known as nitrification as illustrated below

$$NH_3 + 2O_2 \longrightarrow NO_3^- + H_3O^+$$

As potassium dichromate cannot oxidize ammonia to nitrate, the oxygen demand due to nitrification is not included in the *COD* test.

10.6.2 Advantages of *COD* determination over *BOD* determination

(a) In *BOD* determination many oxidisable organic matters do not respond but in *COD* test both biologically oxidisable and biologically inert matters are oxidized.

(b) Determination of *BOD* requires at least 5 days but the determination of *COD* of the same sample requires only 3 hours.

(c) In *BOD* test oxidation process is not at all complete but in *COD* test oxidation process can be assumed to be complete. As *COD* value for a given sample is always higher than *BOD* value of the same sample, *COD* value can be used to estimate approximate *BOD* value.

10.6.3 Nitrogenous *BOD* (*NBOD*)

In any waste water sample, nitrogenous substances remain as ammonia or ammonium salts. This ammonia is oxidised by DO in presence of bacteria according to the following equation

$$NH_3 + 2O_2 \longrightarrow HNO_3 + H_2O$$

$$NH_4^+ + 2O_2 \longrightarrow HNO_3 + H_3O^+$$

1 mole ammonia contains 14 gms nitrogen, which absorbs 64 gms oxygen to form nitric acid. So if nitrogen concentration in a waste water sample is x *mg/l*, the total nitrogenous biochemical oxygen demand (*NBOD*) is $(64/14)x$ *mg/l* or $4.57 \times x$ *mg/l*. However, reaction rate of oxidation of nitrogenous compounds is much slower than that of carbonaceous compounds particularly in presence of bacteria. So, oxidation of nitrogenous compounds occurs only when concentration of carbonaceous compounds is very low or negligible. It is assumed that most of carbonaceous compounds is oxidised within first 5 days and hence presence of nitrogen or nitrogenous substance in the waste does not affect determination of BOD_5.

10.6.3.1 Total Kjeldahl Nitrogen (*TKN*)

For a given sample of waste water, plant effluent or soil, the total amount of nitrogen present in the form of organic nitrogen or ammonia or ammonium salts or a mixture of all three, is called total nitrogen. This total nitrogen is determined by Kjeldahl method. Hence, it is often called total Kjeldahl nitrogen or *TKN*.

10.6.3.2 Kjeldahl method

The Kjeldahl method is schematically shown in the following figure, where, the given waste is heated to a temperature of around 170ºC with H_2SO_4 in presence of catalyst HgO or $CuSO_4$ for an hour to decompose the nitrogenous substance to $(NH_4)_2SO_4$ completely. Then K_2SO_4 is added to increase the temperature of the sample to 189ºC. At this high temperature dark colour of the sample becomes clear and colourless, indicating completion of chemical decomposition of the sample. The solution is then distilled with a little amount of sodium hydroxide to convert ammonium salt to ammonia, which is then absorbed into a solution of boric acid (H_3BO_3). So strength of boric acid solution is reduced and estimated by titrating with standard Na_2CO_3 solution using methyl red indicator.

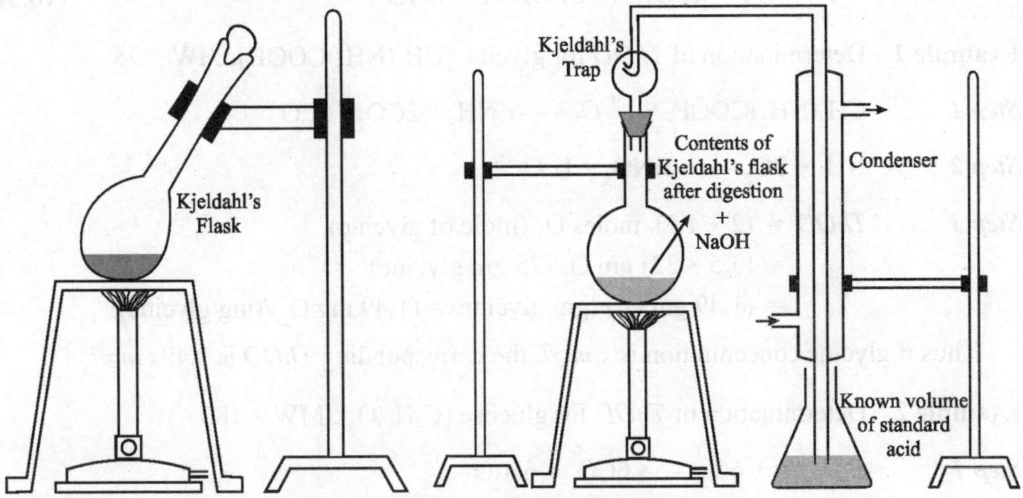

Fig. 10.2 Kjeldahl method to determine *TKN*.

The following reactions may occur during the whole process

$$\text{Waste sample} + H_2SO_4 \longrightarrow (NH_4)_2SO_4(aq) + CO_2(g) + SO_2(g) + H_2O(g)$$

$$(NH_4)_2SO_4(aq) + 2NaOH \longrightarrow Na_2SO_4(aq) + 2H_2O(l) + 2NH_3(g)$$

$$NH_3(g) + H_2O + H_3BO_3 \longrightarrow NH_4^+ + B(OH)_4^-$$

H_3BO_3 is titrated with standard Na_2CO_3 solution

$$H_3BO_3 + H_2O + Na_2CO_3 \longrightarrow NaHCO_3(aq) + NaB(OH)_4(aq) + CO_2(g) + H_2O$$

10.6.3.3 Applications

The method is precise and the results are reproducible and hence it is internationally accepted method to estimate protein content in food samples. However, it cannot measure protein content in a food sample directly because both nonprotein nitrogen and proteinous nitrogen are measured by this method.

10.6.3.4 Disadvantages

- As amino acid sequences are different for different proteins, different correction factors are required.
- Concentrated sulfuric acid is used at high temperature for long time (at least for one hour).

10.6.4 Theoretical oxygen demand (*ThOD*)

For any given waste water sample, *ThOD* is the total amount of oxygen (calculated from stoichiometrically balanced equation) required to oxidise carbon to CO_2, hydrogen to H_2O and nitrogen to nitrates. Thus we can write that

$$ThOD = CBOD + NBOD \quad (10.5)$$

Example 1 Determination of *ThOD* for glycine [$CH_2(NH_2)COOH$], MW = 75

Step 1 $CH_2(NH_2)COOH + 1\frac{1}{2}O_2 \longrightarrow NH_3 + 2CO_2 + H_2O$

Step 2 $NH_3 + 2O_2 \longrightarrow HNO_3 + H_2O$

Step 3 $ThOD = (2 + 1.5)$ moles O_2/(mole of glycine)
$= (3.5 \times 32)$ gm O_2/(75 gm glycine)
$= (1.49$ gm O_2)/(gm glycine) = (1.49 mg O_2)/(mg glycine).

Thus if glycine concentration is *x mg/l*, the corresponding *ThOD* is 1.49*x mg/l*.

Example 2 Determination of *ThOD* for glucose ($C_6H_{12}O_6$), MW = 180

Step 1 $C_6H_{12}O_6 + 6O_2 \longrightarrow 6CO_2 + 6H_2O$

Step 2 $ThOD$ = (6 moles O_2)/(mole of glucose) = (192 gm O_2)/(180 gm glucose)
= (1.07 gm O_2)/(gm glucose) = (1.07 mg O_2)/(mg glucose)

Thus if glucose concentration is *x mg/l*, the corresponding *ThOD* is 1.07*x mg/l*.

Example 3 Determination of *ThOD* for acetic acid (CH_3COOH), MW = 60

Step 1 $CH_3COOH + 2O_2 \longrightarrow 2CO_2 + 2H_2O$

Step 2 $ThOD$ = (2 moles O_2)/(mole of acetic acid) = (64 gm O_2)/(60 gm glucose)
= (1.07 gm O_2)/(gm glucose) = (1.07 mg O_2)/(mg acetic acid)

Thus if acetic acid concentration is *x mg/l*, the corresponding *ThOD* is 1.07*x mg/l*.

10.7 DISSOLVED OXYGEN (*DO*)

Amount of oxygen that remains dissolved in a given water reservoir is called dissolved oxygen or *DO*. It is expressed in *mg/l*. Saturated value of *DO* is around 15 *mg/l* and the critical value of *DO* is around 4 *mg/l*. Critical *DO* value is the minimum *DO* required for survival of aquatic life.

10.7.1 Effects of low *DO* value

(a) Whenever *DO* level drops near or below its critical value, aquatic life becomes endangered.
(b) As *DO* level drops, anaerobic condition prevails, giving rise to floating sludge, bubbling of odorous gases, abnormal fungal growth etc. Ultimately that aquatic ecosystem is destroyed.

10.7.2 Factors influencing *DO* level of a water reservoir

(a) Aquatic plants add *DO* in daytime through photosynthesis but remove *DO* at night through respiration.
(b) Microorganisms consume *DO* due to their respiration and biodegradation of organics.
(c) Intermixing of water sources often increases *DO* content of the resultant flow.
(d) Increase in temperature of water resources reduces solubility of oxygen and hence *DO* content decreases.

10.7.3 Calculation of dissolved oxygen

Solubility of gases in water can be calculated by using Henry's law

$$C_{gas} = k_H \times p_{gas} \qquad (10.6) \text{ [At constant temperature]}$$

C_{gas} = Concentration of dissolved gas in water expressed in moles/lit.
p_{gas} = Partial pressure of the gas in atm.
k_H = Henry's constant in moles/(lit-atm). At a given temperature it varies gas to gas and for a given gas, it varies with temperature as shown below

Table 10.1 k_H value of CO_2 and O_2 at different temperatures.

Temperature (ºC)	CO_2	O_2
0	0.076425	0.0021812
5	0.063532	0.0019126
10	0.053270	0.0016963
15	0.045463	0.0015326
20	0.039172	0.0013840
25	0.033363	0.0012630

In case of oxygen
$C_{gas} = C_{O_2}$ = Equilibrium concentration of oxygen in water = Dissolved oxygen.

$$P_{gas} = p_{O_2}$$

Atmospheric pressure is maximum at the sea level and its value is (P_0). However, this pressure decreases with altitude according to the following relation

$P = P_0 - 1.15 \times 10^{-4} H$, where, H is the altitude in metre. P and P_0 are expressed in atm. As P decreases with altitude, p_{O_2} and hence C_{O_2} also decreases.

10.7.4 Oxygen sag curve

The rate of oxygen consumption or deoxygenation rate directly depends on the concentration of dissolved oxygen present in the system.

Thus, $$-\frac{d[O_2]}{dt} \propto L_t.$$

Or $$-\frac{dO_2}{dt} = k_d L_t$$

Putting the value of L_t from the equation (10.3), we have

$$-\frac{dO_2}{dt} = k_d L_0 e^{-k_d t}$$

Again, the rate of oxygen addition or reaeration rate directly depends on the *oxygen deficit* (D), which is the difference between saturated DO or DO_s and actual DO. So $D = DO_s - DO$. Thus reaeration rate is given by

$$+\frac{dO_2}{dt} \propto D,$$

where, D is the oxygen deficit.

Or $$+\frac{dO_2}{dt} = k_r D$$

Where, k_r is called reaeration rate constant.

The oxygen deficit (D) increases due to deoxygenation but decreases due to reaeration. So, net increase in oxygen deficit with time is given by

$$\frac{dD}{dt} = k_d L_0 e^{-k_d t} - k_r D \tag{10.7}$$

Put, $D = y$. The above equation becomes

$$\frac{dy}{dt} = k_d L_0 e^{-k_d t} - k_r y \tag{10.8}$$

The solution of the above equation is $y = C.F. + P.I.$ (10.9)

$C.F.$ = Complementary Function and

$P.I.$ = Particular Integral.

The equation of *C.F.* is

$$\frac{dy}{dt} + k_r y = 0. \quad \text{Put } y = e^{mt}. \quad \text{Hence, } \frac{dy}{dt} = me^{mt}$$

Putting the values of y and $\frac{dy}{dt}$ in the above equation, we have $m = -k_r$.

$C.F. = Ce^{-k_r t}$, where, C is a constant.

To find out *P.I.*, equation (10.8) is rearranged as follows

$$\frac{dy}{dt} + k_r y = k_d L_0 e^{-k_d t}.$$

Or $$(D + k_r)y = k_d L_0 e^{-k_d t}.$$

where, $$D = \frac{d}{dt}$$

Hence, $$P.I. = \frac{1}{(D+k_r)} k_d L_0 e^{-k_d t} = k_d L_0 \frac{1}{(D+k_r)} e^{-k_d t} = k_d L_0 e^{-k_d t} \frac{1}{(-k_d + k_r)}$$

Or, $$P.I. = k_d L_0 e^{-k_d t} \frac{1}{(-k_d + k_r)}$$

Thus, the solution of equation (10.9) can be written as

$$y = Ce^{-k_r t} + k_d L_0 e^{-k_d t} \frac{1}{(-k_d + k_r)} \tag{10.10}$$

Substituting y by the oxygen deficit D, we have

$$D = Ce^{-k_r t} + k_d L_0 e^{-k_d t} \frac{1}{(-k_d + k_r)} \tag{10.11}$$

Boundary condition: at $t = 0, D = D_0$.

Using this boundary condition in the above equation (10.11), we have

$$C = D_0 - \frac{k_d L_0}{k_r - k_d}$$

Putting the value of C in the equation (10.7), we get:

$$D = \frac{k_d L_0}{k_r - k_d}\left(e^{-k_d t} - e^{-k_r t}\right) + D_0 e^{-k_r t} \tag{10.12}$$

The plot of D versus t curve is shown in the Fig. 10.3.

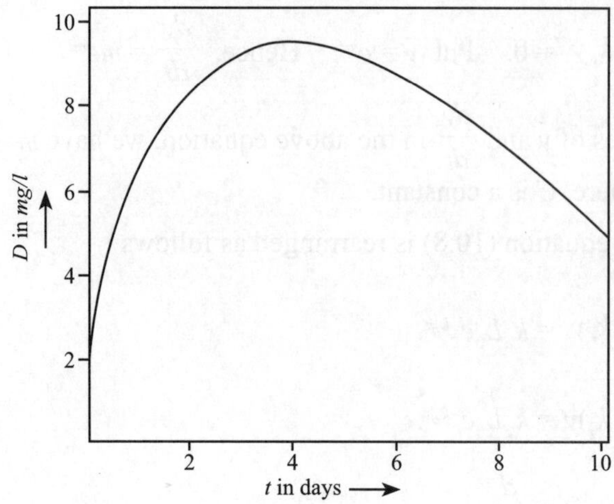

Fig. 10.3 Variation of oxygen deficit (D) with time (t).

The curve shows a maximum at the point O, at which deoxygenation rate is exactly balanced by reaeration rate. The corresponding time is called *critical time*, represented by t_r. To find out the value of t_r the 1st derivative of the equation (10.12) is set to zero. So, we can write that

$$\frac{dD}{dt} = \frac{k_d L_0}{(k_r - k_d)}\left(-k_d e^{-k_d t} + k_r e^{-k_r t}\right) - D_0 k_r e^{-k_r t} = 0 \quad \text{at } t = t_r$$

Dividing both sides by $k_r e^{-k_r t}$ and rearranging we get

$$\left[1 - \frac{k_d}{k_r} e^{(k_r - k_d)t_r}\right] = \frac{D_0 (k_r - k_d)}{L_0 k_d}$$

Hence,

$$t_r = \frac{1}{(k_r - k_d)} \ln\left[\frac{k_r}{k_d}\left(1 - \frac{D_0 (k_r - k_d)}{L_0 k_d}\right)\right] \tag{10.13}$$

10.8 WATER POLLUTANTS

Following are the major water pollutants

(a) Oxygen demanding wastes.
(b) Excess nutrients.
(c) Total dissolved solids.
(d) Heavy metals.
(e) Pesticides.
(f) Volatile organic compounds.
(g) Pathogens.
(h) Thermal pollution.

(a) **Oxygen Demanding Wastes** Organics, present in the water, are oxidized by dissolved oxygen, thereby reducing the dissolved oxygen (*DO*) content. As *DO* level drops, aquatic life is in danger and undesirable colours, odours and tastes are developed in water, reducing its acceptability as a domestic supply. The saturated value of *DO* is DO_s, which is around 14–15 *mg/l*, whereas the critical *DO* level is only 4 *mg/l*.

(b) **Excess Nutrients** Major nutrients for survival of life are nitrogen, phosphorus, carbon, sulphur, Ca, K, Fe, Mn, B, Co etc. However, excess nutrients stimulate the growth of algae, the phenomenon is called **eutrophication**. Whenever the eutrophication starts, reservoir becomes shallower, warmer and is ultimately converted to a marsh. Thus excess nutrients must be avoided as far as practicable.

(c) **Total Dissolved Solids (TDS)** Concentration of total dissolved solids in a given sample of water is known as **salinity** of the sample. It mainly contains some acid radicals and basic radicals.

Acid radicals: Cl^-, SO_4^{2-}, HCO_3^-, NO_3^- etc.

Basic radicals: Na^+, K^+, Mg^{2+}, Ca^{2+} etc.

In fresh water TDS value usually lies within 1500 *mg/l*, in brackish water TDS value usually lies within 5000 *mg/l*, in saline water TDS value is more than 5000 *mg/l*, in sea water TDS value usually lies within 34000 *mg/l* and in drinking water TDS value must lie within 500 *mg/l*.

(d) **Heavy Metals** In chemistry metals, having specific gravity greater than 4 or 5, are known as heavy metals. However, in environmental science, toxic metals are usually called heavy metals. Examples are: As, Cd, Pb, Hg (deadly poisonous), Cu, Mn, Ni, Cr and Fe. Chromium and iron are the essential nutrients for human being but they are extremely toxic if used in higher doses.

(e) **Pesticides** These are often used in agriculture to kill insects. The scientist Müller first invented dichlorodiphenyltrichloroethane (DDT). He was awarded Nobel prize for that in 1948. Other pesticides are benzene hexachloride (BHC), dichlorodiphenyldichloroethene, heptachlor, aldrine, dieldrin, endrin etc. DDT is soluble in lipids and easily get accumulated in fatty tissue. Pesticides are stable for a long period of time in the environment and have bad impact on food chains.

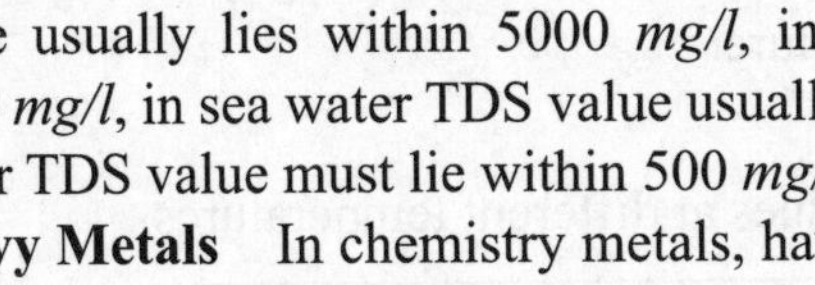

DDT

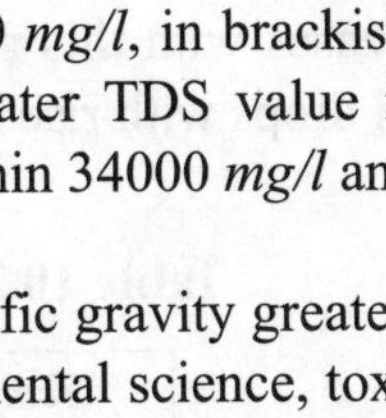

BHC

(f) **Volatile Organic Compounds (VOC)** There are several volatile organic compounds, found partly soluble in water. Out of them five are extremely toxic

(i) *Vinyl chloride or chloroethene* It is the raw material to manufacture Polyvinyl Chloride or PVC. It is a proven carcinogenic compound to human being.

(ii) *Tetrachloroethylene* It is mainly used in heat transfer medium and also in the manufacture of CFC.

(iii) *Trichloroethylene (TCE)* It is commonly used to clean everything from electronic parts to jet engines as well as septic tanks.

(iv) *1,2-Dichloroethane* It is commonly used in the manufacture of vinyl chloride, tetraethyl lead etc. It causes injury to the central nervous system, liver and kidneys.

(v) *Carbon tetrachloride* It is a common household cleaning agent, fire extinguisher and solvent. It is very toxic and also a carcinogenic compound.

(g) **Pathogens** Water-borne disease producing bacteria are called pathogens or coliform bacteria. They are responsible for cholera, dysentery, typhoid etc. Some viruses are responsible for hepatitis. Sewage water contains approximately 500 viruses and 46×10^6 coliform bacteria in 100 ml of sample. Polluted surface water contains 1 virus and 5×10^4 coliform bacteria in 100 ml of sample.

(h) **Thermal Pollution** Several thermal power plants often release huge amount of heat in the nearby river or lake, resulting in sharp rise in temperature. So *DO* level decreases with this rise in temperature. The following chart illustrates how *DO* level drops with rise in temperature.

Table 10.2 *DO* values at different temperatures.

Temperature (°C)	*DO level in mg/l*
0	14.6
10	11.6
20	9.1
30	7.6

10.9 TURBIDITY AND ODOUR

10.9.1 Turbidity

Turbidity of a water sample indicates the haziness. It is measured with the help of an equipment, known as Nephelometer. Hence turbidity is expressed in Nephelometer Turbidity Unit or NTU. The measurement is carried out with respect to a reference sample, which is basically a mixture of hydrazine sulphate and hexamethylenetetramine.

Preparation of the reference sample 1 gm hydrazine sulphate is mixed with 10 gm hexamethylene tetramine in 100 ml distilled water. The turbidity of that reference sample is assumed 400 NTU.

10.9.2 Odour

It refers to foul smell, generated from a given water sample. It is usually expressed in threshold odour number, TON.

Threshold Odour Number (TON) It is the minimum quantity of distilled water, which is to be added to a given waste sample such that odour of the waste sample is just detected. In other words, it is the minimum quantity of distilled water required to ward off any odour from a given waste sample. Mathematically, it can be written as

$$TON = \frac{a+b}{a} \tag{10.14}$$

a = Volume of waste water sample taken. b = Volume of water added.

Significance Less the value of TON indicates less the odour intensity and hence better the quality of waste water sample is.

10.10 DRINKING WATER QUALITY STANDARD

Considering the above water pollutants and their adverse effects on human health, the **World Health Organization (WHO)** has prescribed the following water quality standard.

Table 10.3 Drinking water quality standard.

Impurities	*Concentration*
Zn	5000 *ppb*
Cu	1000 *ppb*
Fe	200 *ppb*
Mn	50 *ppb*
As	50 *ppb*
Pb	50 *ppb*
Cd	10 *ppb*
Hg	1 *ppb*
Chlorides	< 250 *ppm*
Sulphates	< 250 *ppm*
TDS	< 500 *ppm*
pH	7 to 8.5
Turbidity	5 to 10 NTU
Odour	3 TON
Colour	< 10 cobalt scale
Coliform bacteria	0 per 100 ml

10.11 MUNICIPAL WASTE WATER TREATMENT

The waste water plant or effluent treatment plant may be divided into three major sections as given below

(i) Primary treatment plant.
(ii) Secondary treatment plant.
(iii) Tertiary treatment plant.

The overall water treatment process is shown in the Fig. 10.4.

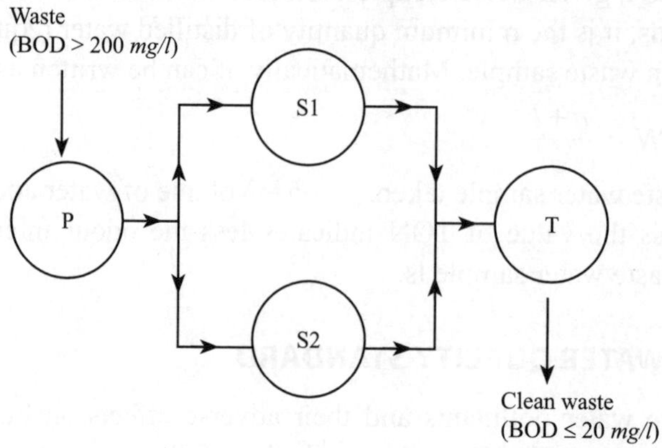

Fig. 10.4 Waste water treatment plant. P = Primary treatment plant. S1 = Trickling filter method. S2 = RBC/Activated sludge method. T = Tertiary treatment plant.

10.11.1 Primary waste water treatment plant

This section consists of two chambers: (a) screening chamber and (b) sedimentation chamber. Raw waste water is first sent to the screening chamber, consisting of several screens of different mesh sizes. While passing through these screens large particles are separated. The process is shown in the Fig. 10.5.

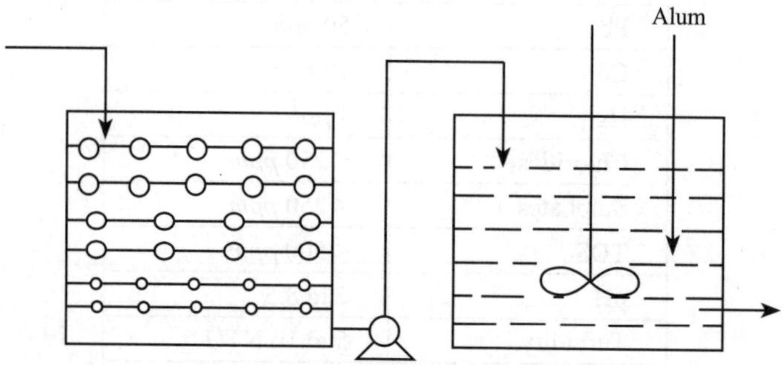

Fig. 10.5 Primary treatment plant, consisting of screening chamber and sedimentation chamber.

The effluent of screening chamber is then sent to the sedimentation chamber where it is treated with alum, namely, $Al_2(SO_4)_3$ or $FeCl_3$ or $FeSO_4$ under anaerobic condition. Alum forms bulky hydroxide precipitates or flocks, floating in the solution. These flocks absorb small colloidal size particles from the effluent and slowly get precipitated at the bottom of the chamber. In presence of air oxidative decomposition sets in, thereby preventing the formation of flocks.

The effluent of sedimentation chamber is then sent to the secondary treatment plant.

10.11.2 Secondary treatment plant

The secondary treatment is carried out by any one of the following methods:

(a) Trickling filter method.
(b) Rotating Biological Contactor (RBC) method.
(c) Activated sludge method.

10.11.2.1 Trickling filter method

The effluent of the primary treatment plant is first sent to a rotary drum with the help of a pump. The effluent is then spread over a filter bed through multiple outlets of the rotary drum. The filter bed is sufficiently porous and aerated so as to allow the effluent flowing through the pores under full aerobic condition. While passing through the pores microbes, distributed in the filter bed, slowly eat up the organics of the effluent. Finally the effluent is taken out from the bottom of the filter bed and sent to a sedimentation chamber where it is allowed to settle for about 12 hours. The heavy sludge is disposed off and the clear effluent is then sent to the tertiary treatment plant. The process is shown in the Fig. 10.6.

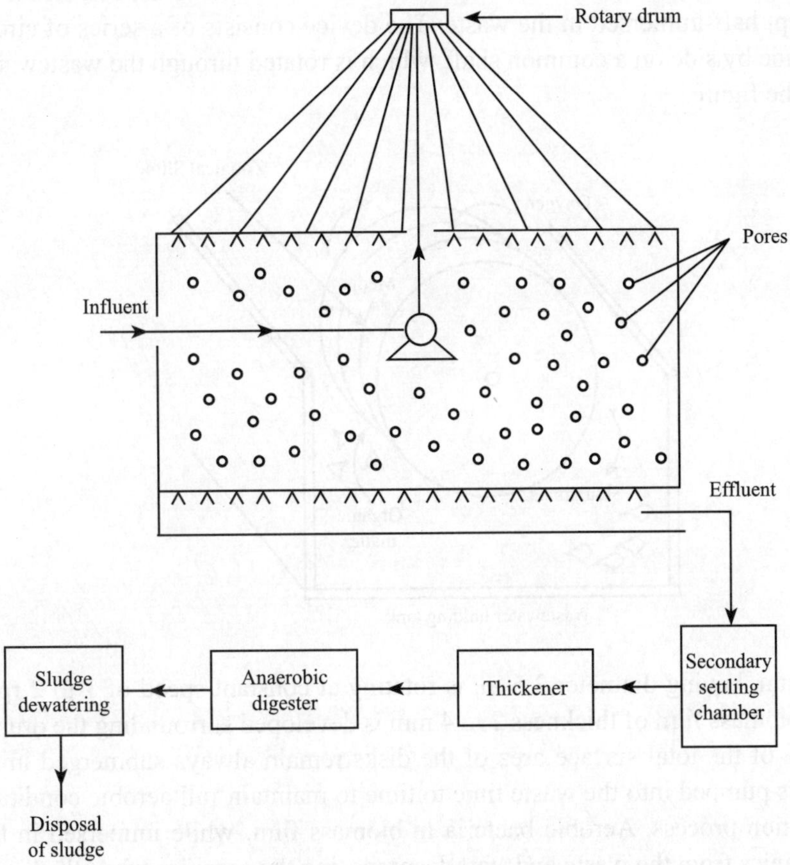

Fig. 10.6 Trickling filter method of secondary treatment plant.

Advantages of trickling filter method

(i) The process is simple and easy to handle.
(ii) It can be easily installed anywhere.
(iii) The process is effective even where waste water contains high concentration of organics and ammonia.
(iv) Low power requirement is another added advantage.

Disadvantages of trickling filter method

(i) The method is costly.
(ii) The biodegradation process is temperature dependent.
(iii) Short filter bed life. When the pores are filled up, the filter bed should be replaced by a new one. Frequent replacement of filter bed decreases efficiency of the method. However, efficiency can be improved by introducing plastic filter media.

10.11.2.2 Rotating Biological Contactor (RBC) method

The effluent of the primary treatment plant is sent to an open chamber, where a rotating drum is kept half-immersed in the waste. The device consists of a series of circular disks mounted side by side on a common shaft, which is rotated through the wastewater flow as shown in the figure.

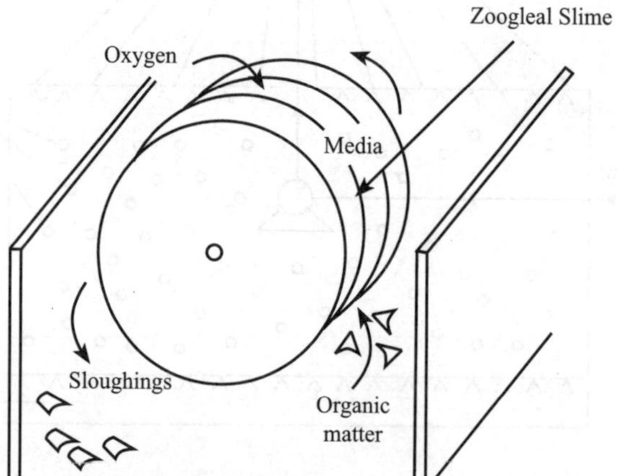

The drum, having diameter 3.6 m, is rotating at constant speed of 1 to 2 rpm. While rotating a biomass film of thickness 2 to 4 mm is developed surrounding the drum surface. About 40% of the total surface area of the disks remain always submerged in the waste water. Air is pumped into the waste time to time to maintain full aerobic condition during biodegradation process. Aerobic bacteria in biomass film, while immersed in the waste, absorb organics from the waste and start decomposing the organics when the biomass film remains in open atmosphere during rotation. The rotating drum is called rotating biological contactor or RBC. An agitator is often attached with the chamber in order to expedite the decomposition process. The process is continued for six to eight hours. After that waste sample is allowed to flow in a settling tank where it is allowed to settle for about 8 hours. Most of the solids are separated as sludge during this period. Clear effluent is taken out from the top and sent to the tertiary treatment plant. The heavy sludge is taken out from the bottom. The sludge is processed and disposed of. A schematic diagram of RBC method is shown in the Fig. 10.7.

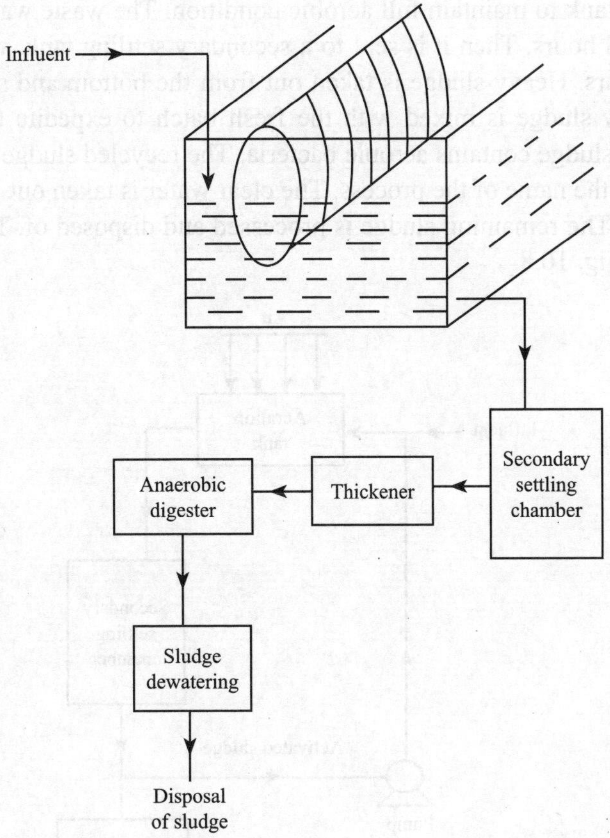

Fig. 10.7 RBC method of secondary treatment plant.

Advantages

(i) As large surface area is available, the process is applicable in wide range of waste-water flow.
(ii) The process is completed in short time period.
(iii) Low power requirement is another advantage.
(iv) High efficiency and excellent process control with low sludge production.

Disadvantages

(i) The thickness of the biomass film increases and hence the RBC needs cleaning regularly.

(ii) Air is often pumped into the biodegradation chamber regularly to maintain full aerobic condition.

10.11.2.3 Activated sludge method

In this method, effluent of primary treatment plant is sent to an aeration tank, where air is pumped to the tank to maintain full aerobic condition. The waste water is continuously agitated for 6 to 8 hours. Then it is sent to a secondary settling tank, where it is allowed to settle for 8 hours. Heavy sludge is taken out from the bottom and sent to a thickener. Part of this heavy sludge is mixed with the fresh batch to expedite the biodegradation process since the sludge contains aerobic bacteria. The recycled sludge is called ***activated sludge*** and hence the name of the process. The clear water is taken out from the top of the settling chamber. The remaining sludge is processed and disposed of. The method is well illustrated in the Fig. 10.8.

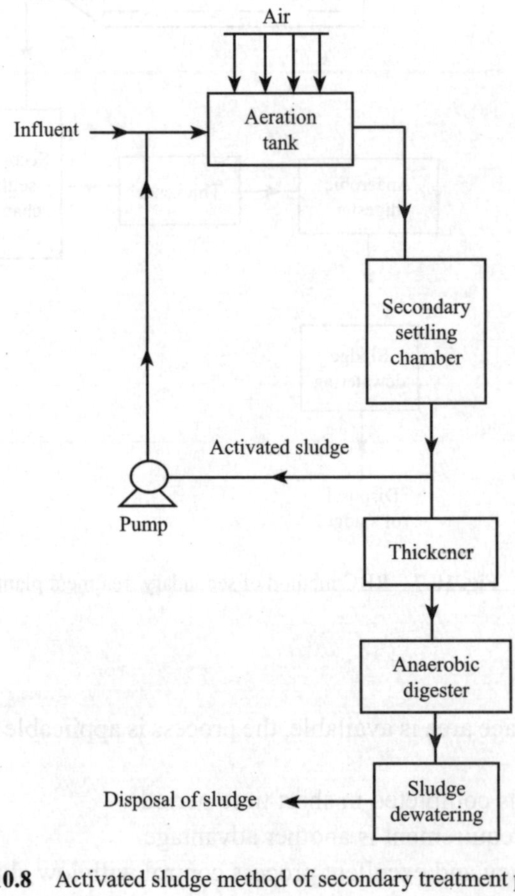

Fig. 10.8 Activated sludge method of secondary treatment plant.

10.11.3 Tertiary treatment plant

The effluent of the secondary treatment plant is first sent to a lime chamber, where the effluent is treated with dry lime (CaO), which reacts with phosphates to form insoluble calcium phosphate [$Ca_3(PO_4)_2$], appearing as fine precipitate at the bottom. Thus phosphates are removed from the effluent.

The clear effluent is then sent to an absorption column from the top. The air is passed from the bottom. While passing through the column the air strips off ammonia from the effluent. The absorption plates are arranged in the column in such a way to have maximum contact between air and effluent. Thus ammonia is removed from the effluent.

The clear effluent is then sent to a disinfectant chamber, where the effluent is treated with bleaching powder or chlorine gas to kill all the bacteria.

Finally the effluent is sent to a pH-controller to maintain pH of the effluent near 8. The acidic waste is treated with slaked lime [$Ca(OH)_2$] and the alkaline waste is treated with carbon dioxide gas to maintain pH near 8. An agitator is attached to the chamber to expedite the process. The clear effluent is then discharged to the nearby reservoir. The tertiary treatment process is shown in the Fig. 10.9.

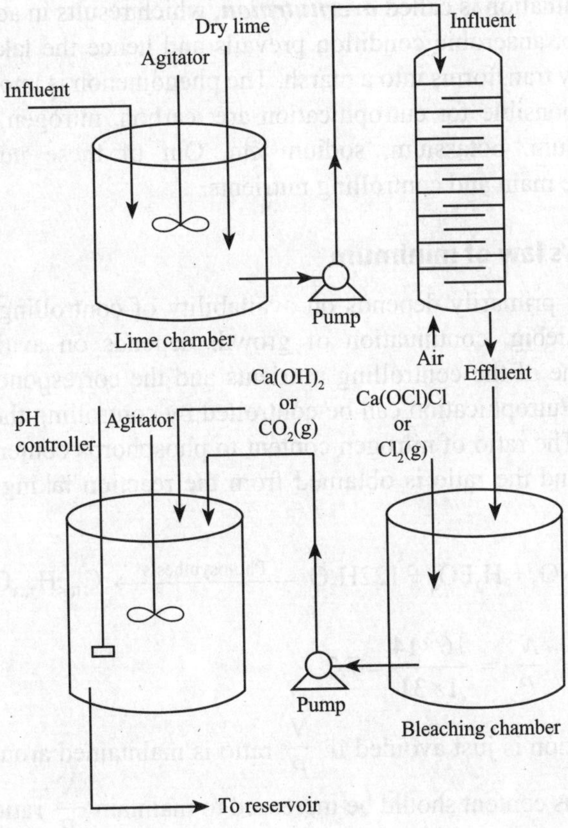

Fig. 10.9 Tertiary treatment plant.

10.12 EUTROPHICATION

10.12.1 Oligotrophic lake

A lake with clear water is called oligotrophic lake. In an oligotrophic lake production of aquatic plants and animals is exactly balanced by its destruction through biodegradation and hence water appears very clear and has sufficient transparency to sunlight. On the basis of penetration of sunlight the oligotrophic lake may be divided into three regions:

(i) Top zone, also known as euphotic zone. This zone receives maximum sunlight and hence contributes the lion's share of aquatic photosynthesis.

(ii) Transition zone, demarcating the top zone and bottom zone. This zone is actually confined to a layer, known as light compensation level, which corresponds to a depth at which the light intensity is about 1% of the full sunlight.

(iii) Bottom zone is known as profundal zone, where photosynthesis hardly occurs.

10.12.2 Overnutrition and Eutrophication

If the effective nutrients in an oligotrophic lake increases abruptly through accumulation of silt and organic matter, the rate of production of aquatic life overshadows the rate of its destruction. The situation is called ***overnutrition***, which results in accumulation of silt and organic debris. So, anaerobic condition prevails and hence the lake becomes shallower, warmer and slowly transforms into a marsh. The phenomenon is known as ***eutrophication***. The nutrients responsible for eutrophication are: carbon, nitrogen, phosphorus, sulphur, calcium, magnesium, potassium, sodium etc. Out of these nutrients, nitrogen and phosphorus are the main and controlling nutrients.

10.12.3 Liebig's law of minimum

Growth of a plant primarily depends on availability of controlling nutrients. According to the scientist Liebig, continuation of growth depends on availability of minimum quantity of any one of the controlling nutrients and the corresponding nutrient is called *limiting nutrient*. Eutrophication can be controlled by controlling the concentration of the limiting nutrient. The ratio of nitrogen content to phosphorus content gives us the idea of limiting nutrient and the ratio is obtained from the reaction taking place during aquatic photosynthesis.

$$106CO_2 + 16HNO_3 + H_3PO_4 + 122H_2O \xrightarrow{\text{Photosynthesis}} C_{106}H_{263}O_{110}N_{16}P + 138O_2\uparrow$$

$$\frac{N}{P} = \frac{16\times 14}{1\times 31} = 7.2$$

So, eutrophication is just avoided if $\frac{N}{P}$ ratio is maintained around 7.0. If $\frac{N}{P}$ ratio is over 10, phosphorus content should be increased to maintain $\frac{N}{P}$ ratio 7.0. So, phosphorus is called limiting nutrient. If $\frac{N}{P}$ ratio is less than 5, nitrogen content should be increased to maintain $\frac{N}{P}$ ratio 7.0. So, then nitrogen is called limiting nutrient. It has been observed that for a given lake, acceptable limits of phosphorus and nitrogen are 0.010 *ppm* and 0.300 *ppm* respectively to avoid any eutrophication.

10.12.4 Causes of eutrophication

(i) Direct discharge of municipal wastes into water reservoir.
(ii) Direct discharge of industrial wastes into water reservoir.
(iii) Fertilizers, pesticides of agricultural lands are carried off by winds and mixed with water.
(iv) Human irresponsibility like direct throwing of food residues into water reservoir.

All the above add nutrients, hence stimulate algal growth and degrade water quality, prevailing anaerobic condition.

10.13 AQUIFERS

10.13.1 Definition and classification

There are several pores in soil and surface water flows down the surface through the pores. This water is stored in different regions inside the Earth's surface as illustrated in the Fig. 10.10.

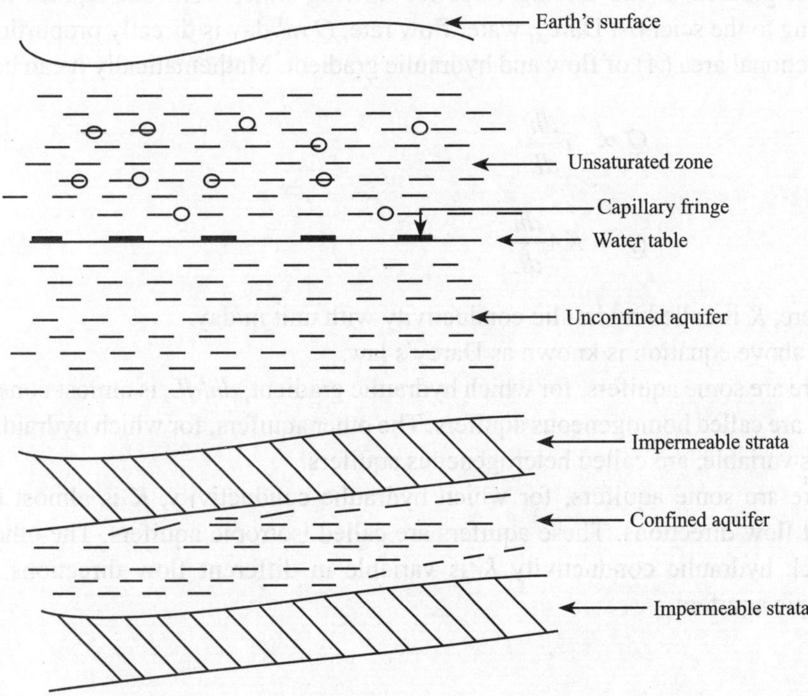

Fig. 10.10 Different water layers below the Earth's surface.

The figure shows that there are three distinct layers of water.

(i) Uppermost layer contains profuse amount of water and air. Water of this layer cannot be pumped out due to the presence of air. This layer of water is called unsaturated zone.

(ii) The layer next to unsaturated zone contains profuse amount of water without any air. Water can easily be pumped out from this layer. This water is called ground water and this ground water resource is called unconfined aquifer. The separation zone between unsaturated zone and unconfined aquifer is called water table, which contains numerous capillary fringe.

(iii) Just below unconfined aquifer impermeable rock is present. Two impermeable rocks are separated by another groundwater resource, known as confined aquifer. Water of this confined aquifer cannot be pumped out due to the presence of impermeable rocks.

Thus aquifers are broadly classified into two categories: (a) confined aquifer and (b) unconfined aquifer.

10.13.2 Darcy's law

In any given area, h represents the distance between water table and Earth's surface and L represents the length along the Earth's surface. As the ground water content varies along the Earth's surface, h often changes with L and the term dh/dL is called hydraulic gradient. Hydraulic gradient is the driving force for flowing water from one aquifer to the other. According to the scientist Darcy, water flow rate, Q m^3/day is directly proportional to both cross-sectional area (A) of flow and hydraulic gradient. Mathematically it can be written as

$$Q \propto A\frac{dh}{dL}$$

Or,

$$Q = KA\frac{dh}{dL}$$

Where, K is called hydraulic conductivity with unit m/day.

The above equation is known as Darcy's law.

There are some aquifers, for which hydraulic gradient, dh/dL, is almost constant. These aquifers are called homogeneous aquifers. The other aquifers, for which hydraulic gradient, dh/dL, is variable, are called heterogeneous aquifers.

There are some aquifers, for which hydraulic conductivity, K is almost constant in different flow directions. These aquifers are called isotropic aquifers. The other aquifers, for which hydraulic conductivity K is variable in different flow directions, are called anisotropic aquifers.

10.14 ESTIMATION OF DISSOLVED OXYGEN (WINKLER'S METHOD)

Theory

Manganous sulphate is used to consume dissolved oxygen of the supplied water sample in presence of an alkali. Divalent Mn is oxidized to tetravalent Mn. The following reaction takes place

$$2MnSO_4 + 4NaOH + O_2 \longrightarrow 2MnO(OH)_2(s) \downarrow + 2Na_2SO_4$$

A brown precipitate of $MnO(OH)_2$ is formed. Instead of pure NaOH solution, alkaline potassium iodide-sodium azide solution is used. The function of sodium azide is to avoid the interference of other oxidizing agents like NO_2^- or HNO_2.

$$2NaN_3 + 2HNO_2 \longrightarrow 3N_2 + 2NO + 2NaOH$$

In the 2nd step of reaction, tetravalent Mn oxidizes I^- to I_2 in acid medium and itself gets reduced to divalent Mn. So the solution, containing brown precipitate, is acidified with H_2SO_4 and the brown precipitate is dissolved to form a clear brown solution, containing divalent Mn and I_2. The following reaction is believed to take place

$$2Mn(SO_4)_2 + 4KI \xrightarrow{H^+} 2MnSO_4 + 2K_2SO_4 + 2I_2$$

The iodine, thus produced, can be estimated by using standardized $Na_2S_2O_3$ solution. The following reaction takes place during titration

$$4Na_2S_2O_3 + 2I_2 \longrightarrow 2Na_2S_4O_6 + 4NaI$$

The above three reactions show that 1 mole of dissolved O_2 produces 2 moles of $MnO(OH)_2$, which produces 2 moles of I_2, which consumes 4 moles of $Na_2S_2O_3$. Thus concentration of dissolved oxygen can be directly calculated from the number of moles of thiosulphate consumed.

Apparatus BOD bottle, Burette, pipette, conical flask, volumetric flask, beaker.

Reagents KIO_3/KI solution, $Na_2S_2O_2$ solution, alkaline KI-azide solution, manganous sulphate solution, starch indicator.

Preparation of standard KIO_3/KI solution

Take 180 mg KIO_3 in a 1 litre volumetric flask. Dissolve it in a little amount of distilled water. Add 1.5 gm KI into it. Shake it well till a clear solution is obtained. Fill the volumetric flask up to the mark. Thus 1 litre 0.5056 (N/100) standard solution is prepared.

Preparation of 0.5 (N/10) $Na_2S_2O_3$ solution

Approximately 12.4 gm $Na_2S_2O_3.5H_2O$ crystal along with 1 gm Na_2CO_3 is dissolved in a 1000 ml volumetric flask. Na_2CO_3 acts as a stabilizer of $Na_2S_2O_3$ by making the solution alkaline. Thus 1 litre $Na_2S_2O_3$ solution of approximate strength 0.5 (N/10) is prepared.

Preparation of alkaline KI-azide solution

5 gm of KI is dissolved in 100 ml distilled water. 5 gm NaOH is added to it. 5 mg sodium azide is added to it. Thus alkaline KI-azide solution is prepared.

Preparation of manganous sulphate solution

10 gm $MnSO_4.4H_2O$ is dissolved in 100 ml distilled water. So approximately 10% manganous sulphate solution is prepared.

Procedure

Standard KIO_3/KI solution of strength 0.5056 (N/100) is supplied. 1.5 N H_2SO_4 is supplied.

10 ml supplied 0.5 (N/10) $Na_2S_2O_3$ solution is taken in a 100 ml volumetric flask. It is then filled up to the mark by distilled water. Thus $Na_2S_2O_3$ solution of approximate strength 0.5 (N/100) is prepared.

Rinse burette, pipette and conical flask with distilled water. $Na_2S_2O_3$ solution is to be poured in a burette. 10 ml supplied KIO_3/KI solution is to be aliquoted and poured in a 250 ml conical flask. 10 ml distilled water is added to it followed by 1 ml 1.5(N) H_2SO_4. The solution turns deep brown colour. It is then titrated against $Na_2S_2O_3$ solution, running from a burette till the colour fades. Then 1 ml starch solution is added. The mixture turned to deep blue colour. The titration is resumed till the colour disappears permanently. Initial and final readings are noted and three such results are taken.

90 ml sample water is to be taken in a 100 ml volumetric flask. 2 ml $MnSO_4$ solution is also added to it by using measuring cylinder. 2 ml alkaline KI-azide solution is added to it by using measuring cylinder. The bottle is to be stoppered properly. Brown precipitate of $MnO(OH)_2$ appears and slowly settles down at the bottom. 5 ml 1.5(N) H_2SO_4 is quickly added to it by using measuring cylinder and the volumetric flask is stoppered immediately. The volumetric flask is to be shaken to dissolve out the precipitate. The solution turns to deep brown colour. 25 ml of the solution is aliquoted and transferred to a 250 ml conical flask. It is then titrated against standardized sodium thiosulphate solution till the colour fades. Then 1 ml starch solution is added. The mixture turns to deep blue colour. The titration is resumed till the colour disappears permanently. Initial and final readings are noted and three such results are taken.

Results

Table 10.4 Standardization of $Na_2S_2O_3$ solution.

No. of observations	*Volume of KIO_3/KI solution taken (ml)*	*Volume of $Na_2S_2O_3$ solution consumed (ml)*			*Mean volume (ml)*
		Initial	*Final*	*Difference*	
1	10				
2	10				
3	10				

Calculation

$V_1S_1 = V_2S_2$

V_1 = Volume of KIO_3/KI solution taken = 10 ml

S_1 = Strength of KIO_3/KI solution = 0.5056 (N/100)

V_2 = Mean burette reading = … … ml

S_2 = Strength of $Na_2S_2O_3$ solution = y(N/100).

Table 10.5 Estimation of dissolved oxygen.

No. of observations	*Volume of sample water taken (ml)*	*Volume of $Na_2S_2O_3$ solution consumed (ml)*			*Mean volume (ml)*
		Initial	*Final*	*Difference*	
1	25				
2	25				
3	25				

Calculation

$V_1S_1 = V_2S_2$

V_1 = Volume of sample water solution taken = 25 ml

S_1 = Strength of sample water solution with respect to dissolved oxygen = f(N).

V_2 = Mean burette reading = … …ml

S_2 = Strength of $Na_2S_2O_3$ solution = y(N/100) = (y/100) N = known from Table 1.

Equivalent weight of atomic oxygen is 8 since atomic oxygen accepts 2 electrons to form O^{2-}. Thus,

S_1 = Strength of sample water solution = f(N) = f (gm – eq.)/litre = $f \times 8$ (gm oxygen)/litre.

So, amount of dissolved oxygen = 8 f *gm/l* = 8000 f *mg/l* = 8000 f *ppm*.

Conclusion

Amount of dissolved oxygen in the supplied water sample is … … … … *ppm*.

10.15 NUMERICAL PROBLEMS

Problem 1 The dissolved oxygen in an unseeded sample of diluted waste having an initial *DO* of 9 *mg/l* is measured to be 3 *mg/l* after 5 days. The dilution factor (P) is 0.030 and the reaction rate constant k is 0.22 day^{-1}.

(a) What is the 5-day *BOD* of the waste?

(b) What would be the ultimate carbonaceous *BOD*?

(c) What would be the remaining oxygen demand after 5 days?

Solution $BOD_5 = L_0\,(1 - e^{-k_d t})$. P = 0.030.

$$BOD_5 = \frac{DO_i - DO_f}{P} = \frac{9-3}{0.03} = 200\ mg/l$$

$t = 5$ days, $k_d = 0.22$ day^{-1}. Putting these values in the above equation, we get

$$L_0 = \frac{200}{1-e^{-0.22\times 5}} = 299.79\ mg/l.$$

$$L_t = L_0 - BOD_5 = (299.79 - 200)\ mg/l = 99.79\ mg/l$$

Problem 2 A test tube containing just seeded dilution water has its *DO* level drop by 1 *mg/l* in a 5-day test. In a 300 ml *BOD* bottle filled with 15 ml of waste water and the rest seeded dilution water (dilution 1:20) experience a drop of 7.2 *mg/l* in the same time period. What would be the 5-day *BOD* of the waste?

Solution Dilution factor = P = 15/300 = 0.05.

$$BOD_5 = \frac{(DO_i - DO_f) - (DW_i - DW_f)(1-P)}{P} = \frac{BOD_m - BOD_w(1-P)}{P}$$

$BOD_m = 7.2\ mg/l$, $BOD_w = 1\ mg/l$

$$BOD_5 = \frac{7.2 - 1(1-0.05)}{0.05}\ mg/l = 125\ mg/l$$

Problem 3 A waste had an ultimate *BOD* equal to 300 *mg/l*. At 20ºC the 5-day *BOD* was 200 *mg/l*. What would be the 5-day *BOD* of the waste at 25ºC?

Solution $k_\theta(k_d \text{ at } t^\circ C) = k_{20}(k_d \text{ at } 20^\circ C)\alpha^{(\theta-20)}$. $\alpha = 1.047$

At 20ºC, $L_0 = 300\ mg/l$, $BOD_5 = 200\ mg/l$. $200 = 300(1 - e^{-k_{20}\times 5})$

Hence, $k_{20} = 0.22$ day^{-1}. Thus, $k_{25} = 0.22 \times 1.047^{(25-20)} = 0.277$ day^{-1}

At 25ºC $BOD_5 = 300(1 - e^{-k_{25}\times 5}) = 300(1 - e^{-0.277\times 5}) = 225\ mg/l$

Problem 4 A municipal waste water treatment plant, serving a city of 2×10^5, discharges 1.10 m^3/s of treated effluent, having an ultimate *BOD* of 50 *mg/l* into a stream that has a flow of 8.70 m^3/s and an ultimate *BOD* of its own equal to 6 *mg/l*. The deoxygenation rate constant k_d is 0.20 day^{-1}.

(a) Assuming complete and instantaneous mixing, estimate the ultimate *BOD* of the river just downstream from the outfall.

(b) If the stream has constant cross-section so that it flows at a fixed speed equal to 0.30 m/s, estimate the *BOD* of the stream at a distance 30 km downstream.

Solution

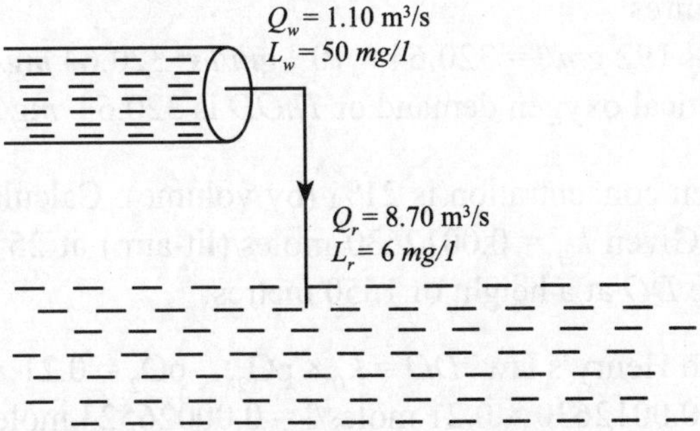

Q_w = Flow rate of waste water = 1.10 m³/s,

L_w = Ultimate *BOD* of waste water = 50 *mg/l*.

Q_r = Flow rate of the stream = 8.70 m³/s,

L_w = Ultimate *BOD* of the stream = 6.0 *mg/l*.

Thus L_0 at the mixing point is given by

$$L_0 = \frac{Q_w L_w + Q_r L_r}{Q_w + Q_r} = \frac{1.10 \times 50 + 8.70 \times 6}{1.10 + 8.70} = 10.94\ mg/l$$

Thus the ultimate *BOD* of the river just downstream from the outfall is 10.94 *mg/l*.

$v = 0.30$ m/s, $d = 30000$ m, $t = 30000/(0.30 \times 86400)$ day = 1.157 day

So, time taken to cover 30 km distance is 1.157 day.

We know: $BOD_5 = L_0(1 - e^{-k_d t})$

$BOD_5 = L_0(1 - e^{-k_d t}) = 10.94\ (1 - e^{-0.20 \times 1.157}) = 2.26\ mg/l.$

Problem 5 A waste water sample has BOD_5 equal to 180 *mg/l* and the reaction rate constant equal to 0.22 day^{-1}. It has total nitrogen content of 30 *mg/l*.

(a) Find the ultimate *CBOD*.

(b) Find the ultimate *NBOD*.

(c) Find the remaining *BOD* after 5 days.

Solution $BOD_5 = L_0(1 - e^{-k_d t})$. $BOD_5 = 180\ mg/l$. $k_d = 0.22$ day^{-1}

$180 = L_0(1 - e^{-0.22 \times 5})$. $\therefore\ L_0 = 269.81\ mg/l.$

$NBOD = 4.6 \times 30\ mg/l = 138\ mg/l.$

Problem 6 1.67×10^{-3} M glucose ($C_6H_{12}O_6$) solution is completely oxidized to CO_2 and H_2O. Calculate the theoretical oxygen demand.

Solution Oxidation of glucose occurs according to the following equation

$$C_6H_{12}O_6 + O_2 \longrightarrow 6CO_2 + 6H_2O$$

Thus 1 mole glucose requires 6 moles or 192 gm oxygen. Thus 1.67×10^{-3} M glucose requires
$1.67 \times 10^{-3} \times 192$ *gm/l* = 320.64×10^{-3} *gm/l* = 320.64 *mg/l* oxygen.
Thus theoretical oxygen demand or *ThOD* is 320.64 *mg/l*.

Problem 7 In air oxygen concentration is 21% (by volume). Calculate the *DO* at 25ºC and 1 atm. Given k_H = 0.0012630 moles/(lit-atm) at 25ºC and 1 atm. Also calculate the *DO* at a height of 1550 metres.

Solution According to Henry's law $DO = k_H \times pO_2$, $pO_2 = 0.21 \times 1$ atm = 0.21 atm.
Thus $DO = 0.0012630 \times 0.21$ moles/l = 0.00026523 moles/l = 8.49 *mg/l*.
At any altitude H the atmospheric pressure is given by $P = P_0 - 1.15 \times 10^{-4}$ H.
Thus atmospheric pressure at a height of 1550 metres is given by
$P = 1 - 1.15 \times 10^{-4} \times 1550 = 0.82175$ atm, where, $P_0 = 1$ atm.
So, $pO_2 = 0.21 \times 0.82175$ atm = 0.173 atm.
Thus $DO = 0.0012630 \times 0.173$ moles/l = 0.0002185 moles/l = 7 *mg/l*.

Exercise 1 A waste water treatment plant discharges 1 m^3/s of effluent having an ultimate *BOD* of 40 *mg/l* into a stream flowing at 10 m^3/s. Just upstream from the discharge point the stream has an ultimate *BOD* of 3 *mg/l*. The deoxygenation rate constant k_d is estimated at 0.22 day^{-1}.
Assuming complete and instantaneous mixing, find the ultimate *BOD* of the mixture of waste and river just downstream from the outfall.

Assuming a constant cross-sectional area for the stream equal to 55 m^2, what would be the ultimate *BOD* at a point 10 km downstream.

Exercise 2 A standard *BOD* test is run using seeded dilution water. In one bottle the waste sample is mixed with seeded dilution water giving dilution of 1:30. Another bottle contains just seeded dilution water. Both bottles begin the test with a *DO* of saturation value of 9.2 *mg/l*. After 5 days the bottle containing waste has a *DO* equal to 2 *mg/l*, while that containing just seeded dilution water has *DO* equal to 8 *mg/l*. Find the 5-day *BOD* of the waste.

Exercise 3 If BOD_5 of a waste water sample is 200 *mg/l* and the ultimate *BOD* is 300 *mg/l*, find the reaction rate constant.

Exercise 4 A waste water sample has 5-day *BOD* at 20ºC equal to 210 *mg/l* and the ultimate *BOD* of 350 *mg/l*. Find the 5-day *BOD* at 25ºC.

Exercise 5 A mixture consisting of 30 ml of waste and 270 ml of seeded dilution water has an initial *DO* of 8.55 *mg/l*. After 5 days it has a final *DO* of 2.40 *mg/l*. Another bottle containing just seeded dilution water has an initial *DO* of 8.75 *mg/l* and a final *DO* of 8.53 *mg/l*. Find the 5-day *BOD* of the waste.

11

Metal and Nuclear Pollutions

11.1 COPPER

Copper is extensively used in household purposes. It is a toxic element and more toxic in ionic form. Around 1.7 *ppm* of copper in drinking water has deleterious effect on health. It causes fatal hemolytic anemia, fatty metamorphosis of liver, narcosis of kidney. Permissible level in potable water is 1 *ppm* and in soil is 0.2 *ppm*. The prescribed TLV is 1 mg/m^3 but 0.2 mg/m^3 is the best for all purposes. ESP and baghouse filter are usually used for collection of copper dust. Wheat flour, animal feed contain considerable amount of copper. Jewellery effluents, copper mines, copper industries are the main sources of copper in the environment.

11.2 TOXIC EFFECTS OF CADMIUM (Cd)

11.2.1 Source

Cladding industry, nuclear reactor, alkaline cells and alloy industries. Substantial amount of cadmium is released in air when rubber tyres and plastic containers are burnt.

11.2.2 General adverse effects

1. Causes high blood pressure.
2. It decomposes RBC. So it is called hematoxin.
3. It damages kidneys. So it is called nephrotoxin.
4. Aquatic life becomes endangered in presence of cadmium.
5. It has affinity to *S* and hence reduces the activity of enzyme by reacting with –*SH* groups of enzymes.
6. Bones become fragile in presence of Cd.
7. It is responsible for cancer. So it is a carcinogenic element.
8. Gastric and intestinal disorder is another adverse effect of Cd.

11.2.3 Biochemical effects

Cadmium, ingested by human being, is normally trapped in kidneys. However, 99% Cd is eliminated while only 1% is stored. If the Cd content in the body is high, more than 1% Cd is stored in kidneys. This excess Cd^{2+} then replaces Zn^{2+} in Zn containing enzymes, resulting metabolic disorder. Like As, Cd also has affinity to *S* and hence attacks *–SH* groups of enzymes, reducing the activity of enzymes. Cadmium is deadly poisonous, responsible for renal disorder, anemia, hypertension and bone marrow disorder.

11.2.4 Preventive measures against cadmium pollution

(i) Locate and seal the cadmium pollution sources.
(ii) Regular monitoring of cadmium pollution level.
(iii) Take necessary remedial actions. Cd-level should not exceed 10 *ppb*. Cd-toxicity can be reduced by using Zn.

11.3 TOXIC EFFECTS OF LEAD (Pb)

11.3.1 Source

Source gasoline, lead pigment industries, ceramic utensils, glass components. Auto-exhaust emissions contain considerable amount of lead.

To reduce the lead content in auto-exhaust emission catalytic converter is used. **Organic lead** causes intoxication, disrupts central nervous system, weakness in wrists and legs, anemia, gastrointestinal disturbance, cardiac disorders. In women it causes infertility while in men it weakens reproductive system. **Inorganic lead** causes abdominal pain, asthma, constipation, diarrhea, vomiting etc.

11.3.2 General adverse effects

1. It causes anemia.
2. It causes kidney failure.
3. It is responsible for nervous disorder.
4. It is responsible for soil pollution.
5. It is one of the components of SPM.

11.3.3 Biochemical effect

Haem is the prosthetic group of hemoglobin, myoglobin, and cytochromes. Haem is an **asymmetric** molecule, containing methyl group, sulphur and protein as shown in the following figure. Condensation of glycine and succinyl-CoA, with decarboxylation, results δ-aminolevulinic acid (ALA), which is ultimately converted to haem.

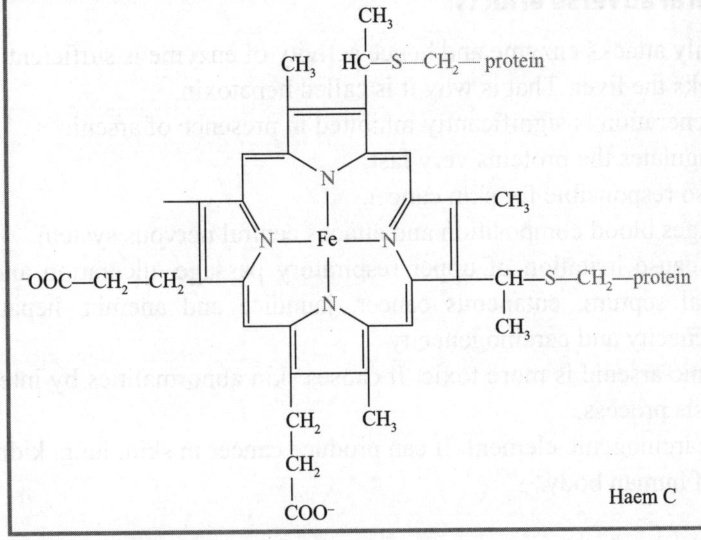
Haem C

Pb inhibits d-aminolevulinic acid to synthesize haem. It also inhibits proteins to carry O_2 and glucose. All these effects are detected when Pb level in the blood reaches about 0.3 *ppm*. When the Pb level in the blood is more than 0.6 *ppm*, the following symptoms are detected: (i) anemia, (ii) renal disorder and (iii) brain damage.

Excess Pb can be removed from human body by using chelating agents, like, EDTA, 2,3-dimercaptopropanol, d-penicillamine etc. Pb readily forms stable chelates with them and is finally removed through urine. In this way, Pb poisoning can be cured.

11.4 CONTROL MEASURES AGAINST LEAD POLLUTION

(i) Lead compounds should be replaced by other harmless compounds.
(ii) Use of unleaded petrol in automobiles is highly recommended.
(iii) The places where lead particles are generated should be cordoned.
(iv) Proper ventilation of exhaust gases, mixed with lead particles, should be arranged.
(v) Gas masks should be used to stop inhalation of gases.
(vi) Factory exhaust should be treated properly before releasing to the atmosphere.

11.5 TOXIC EFFECTS OF ARSENIC

As(III) is the most toxic. It is generally found in insecticides, fungicides and herbicides.

11.5.1 Source

Main sources are: cotton, electronic, photocopying, computer industries, pharmaceutical, paper-pulp and glass industries.

11.5.2 General adverse effects

1. It readily attacks enzyme and hence activity of enzyme is sufficiently reduced.
2. It attacks the liver. That is why it is called hepatoxin.
3. ATP generation is significantly inhibited in presence of arsenic.
4. As coagulates the proteins very fast.
5. It is also responsible for skin cancer.
6. It changes blood composition and attacks central nervous system.
7. It can cause irritation of upper respiratory passage, ulceration and perforation of nasal septum, cutaneous cancer, jaundice and anemia, hepatic disorders, mutagenecity and carcinogenecity.
8. Inorganic arsenic is more toxic. It causes skin abnormalities by interfering DNA synthesis process.
9. It is a carcinogenic element. It can produce cancer in skin, lung, kidney and other parts of human body.

11.5.3 Biochemical effect

As(III) is the most toxic. It has a natural penchant for *S* and hence it readily attacks *–SH* groups of enzyme. So enzyme action in human is sufficiently inhibited in presence of As. The following reaction is believed to occur

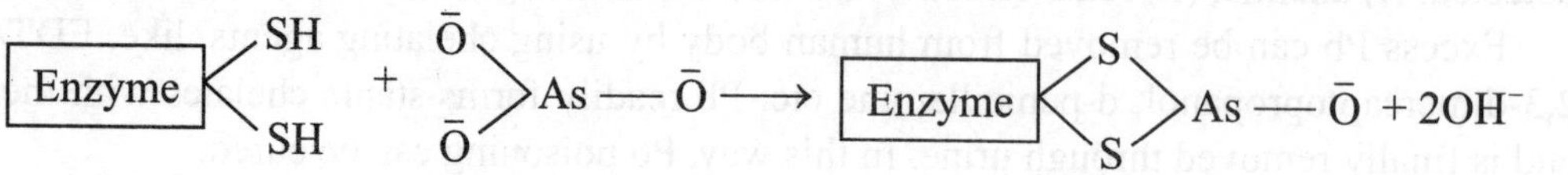

Enzymes generate energy through biochemical process but As in the form of arsenate strongly inhibits the enzymatic action.

Another energy producing substance in human body is ATP (adenosine triphosphate). Glyceraldehyde 3-phosphate is converted to ATP through biochemical process. Arsenic (As) and phosphorus (P) belong to the same group. Thus in arsenic contamination As often replaces P of glyceraldehyde 3-phosphate, thereby inhibiting ATP generation. The following process is believed to take place during ATP generation.

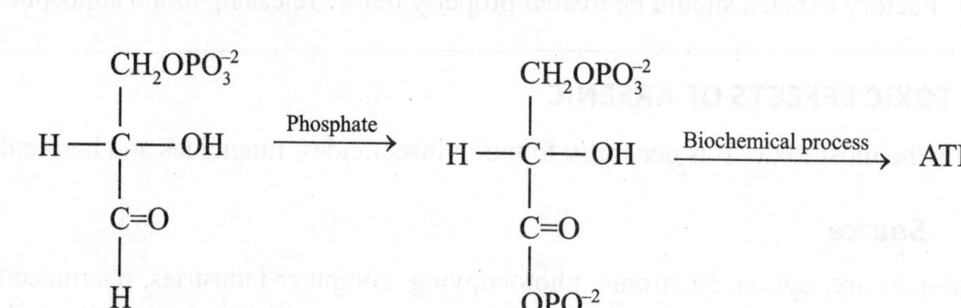

glyceraldehyde 3-phosphate   1,3-diphosphoglycerate

However, in presence of AsO_3^{3-}, 1,3-diphosphoglycerate is not formed preventing ATP generation.

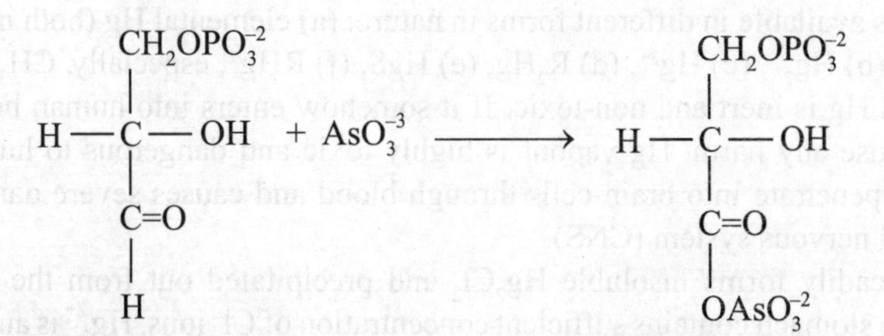

glyceraldehyde 3-phosphate 1-arseno-3-phosphoglycerate

11.5.4 Preventive measures against arsenic pollution

(i) Identify the sources of arsenic pollution and cordon the area.

(ii) In case of arsenic polluted groundwater sources the inhabitants must be prevented from using drinking water from that area. Deeper tube wells (70 metres to 100 metres) should be used for drinking water.

(iii) Regular monitoring of water samples of groundwater and drinking water is necessary to control arsenic level.

(iv) Use of rain water for drinking purposes.

(v) Reducing the use of arsenic enriched insecticides and pesticides.

11.6 TOXIC EFFECTS OF MERCURY

11.6.1 Introduction

It is the king of toxic elements. Liquid mercury has no adverse effect on health but its vapour is deadly toxic. So workers should not be exposed to mercury vapours for more than 30 hours per week. Organic mercury is more dangerous than inorganic mercury. In Minamata (Japan) a paper industry released mercury wastes in sea and these mercury wastes contained $(CH_3)_2Hg$ and $(C_2H_5)_2Hg$. They were consumed by fishes and finally it moved to human body through food chain. Several persons died due to mercury poisoning.

11.6.2 General adverse effects

1. Elemental Hg is non-toxic but Hg vapour is extremely toxic when inhaled.
2. Toxicity level of Hg_2^{2+} is low but that of Hg^{2+} is high and the latter does not have easy access to the biological membranes.
3. RHg^+, organomercurials, is extremely toxic. It is easily transported to biological membranes. It causes nervous breakdown, damages brain cells.
4. RHg^+ decomposes blood cells, attacks haemoglobin, causing anemia. Hence Hg is also called hematoxin.
5. RHg^+ damages kidney. Hence it is also called nephrotoxin.
6. It has affinity to *S* and hence reduces the activity of enzyme by reacting with *–SH* groups of enzymes.

11.6.3 Biochemical effect

Mercury (Hg) is available in different forms in nature: (a) elemental Hg (both in liquid and vapour states), (b) Hg_2^{2+}, (c) Hg^{2+}, (d) R_2Hg, (e) HgS, (f) RHg^+, especially, CH_3Hg^+.

(a) Liquid Hg is inert and non-toxic. If it somehow enters into human body, it does not cause any harm. Hg vapour is highly toxic and dangerous to human being. It can penetrate into brain cells through blood and causes severe damage to the central nervous system (CNS).

(b) Hg_2^{2+} readily forms insoluble Hg_2Cl_2 and precipitated out from the system. As human stomach contains sufficient concentration of Cl^- ions, Hg_2^{2+} is automatically isolated from the system as $Hg_2Cl_2(s)$. Thus Hg_2^{2+} is not so toxic to human being.

(c) Hg^{2+} is highly toxic. It has a strong penchant to *S*– atoms. So it attacks –*SH* groups of enzyme, retarding the enzymatic action significantly. It also attacks *S*– bearing amino acids of proteins, forms bonds with haemoglobin, albumin etc. It is unable to move across biological membranes and hence biological cells are not affected by Hg^{2+}.

(d) Organomercurials, i.e., RHg^{2+} are highly toxic species. Most common and available organomercurial is CH_3Hg^+. It can easily penetrate through biological membranes. It is responsible for creating energy deficiency in brain cells, resulting collapse of central nervous system. CH_3Hg^+ accelerates segregation of chromosome and chromosome breakage in cell. It also inhibits cell division. It is responsible for clotting in blood cell. It also attacks kidney, resulting severe damage to kidney system. Only 0.5 *ppm* of CH_3Hg^+ is enough to execute the above dreadful activities.

(e) R_2Hg has a very low toxicity. However, it can be converted to RHg^+ in acidic medium.

(f) HgS is insoluble and apparently non-toxic.

11.6.4 Preventive measures against mercury pollution

(i) All mercury based pesticides must be banned.

(ii) Use of mercury compounds should be restricted to selected areas only.

(iii) Chlor-Alkali plants must stop using Hg-electrodes.

11.7 NUCLEAR POLLUTION

11.7.1 Radiation measurement

Strength of radioactive radiation is usually expressed in Curie or Ci. However, in SI (System Internationale) unit, it is expressed in Becquerel or Bq. These two units are defined as follows

1 becquerel (Bq) = 1 disintegrations per second or dps.

1 Curie (Ci) = 3.7×10^{10} dps.

1 rad (R) = It is the quantity of γ-radiation, which causes ionisation equivalent to 1 esu of electricity. This unit is widely used.

The effect of radioactive radiation is measured as follows

(i) The radiation dose absorbed by a person. The amount of energy deposited in human tissue due to absorption of radiation dose by a person is expressed either by conventional unit rad or by SI unit gray (Gy). 1 Gy = 100 rad.

(ii) The risk to health that a person will suffer due to exposure to radiation is called biological risk. It is expressed either by conventional unit rem or by SI unit sievert (Sv). Harmful ionizing radiations are α-particles, β-particles, X-rays and γ-rays. Scientists have assigned a number to each type of radiation depending on its degree of harmfulness or energy transfer to cells of the body. This number is known as Quality Factor (Q). On the basis of this Quality Factor, rem is defined as rem = rad × Q and 1 Sv = 100 rem.

International Commission on Radiological Protection (ICRP) has advised a maximum dose of 17 mrem/year for a normal human being. People are exposed to radiation from different sources, such as naturally occurring radioactive materials in the soil and cosmic rays from outer space. Some sources of exposure to radiations and the associated doses are shown in the table below.

Table 11.1 Common radiation exposures.

Source of exposure	*Dose in rem*	*Dose in sievert (Sv)*
Exposure to cosmic rays during travelling in airplane flight	3 mrem	0.03 mSv
One dental X-ray	4–15 mrem	0.04–0.15 mSv
One chest X-ray	10 mrem	0.1 mSv
One mammogram	70 mrem	0.7 mSv
One year of exposure to natural radiation	300 mrem	3 mSv

11.7.2 Classification of nuclear wastes

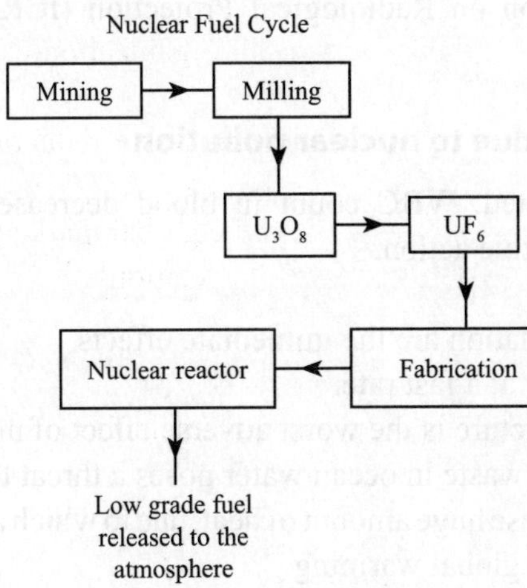

Low grade fuel, released to the atmosphere, is known as nuclear waste. This nuclear waste can be classified into three categories

(a) The nuclear waste, containing low concentration of short lived radio-nuclides, is called ***low level*** nuclear waste.

(b) The nuclear waste, containing high concentration of highly active and long lived radio-nuclides, is called ***high level*** nuclear waste. This type of waste emits heat even after reprocessing.

(c) The ***intermediate level*** nuclear waste contains both short and long lived radio-nuclides in significant amount.

Facts

(a) In 1965 USSR dumped about 2.5 million Curies of radioactive waste into ocean water.

(b) Chernobyl disaster took place in 1986 due to huge release (around 80 million Curies) of Sr^{90}, Cs^{137} and I^{131}.

Extraction and processing of nuclear materials as well as disposal of nuclear waste poses a severe threat on environment and all living beings.

Table 11.2 Dose-Response Chart.

Dose (mille Rem/year)	*Response*
25	Detectable
50	Change in blood
100	Nausea, vomiting, illness
250	Recoverable illness
500	Unrecoverable or difficultly recoverable illness
1000	Certain death

International Commission on Radiological Protection (ICRP) devised the standard dose of 17 mille Rem/year.

11.7.3 Adverse effects due to nuclear pollution

(a) Blood cell is affected. WBC count in blood decreases, thereby reducing the defence and preventive action.

(b) Intestine is affected.

(c) Cancer, tumour, epilation are the immediate effects.

(d) Loss of biodiversity at a fast rate.

(e) Change in gene structure is the worst adverse effect of nuclear pollution.

(f) Disposal of nuclear waste in ocean water poses a threat to aquatic life.

(g) Radio-nuclides release huge amount of heat, due to which atmospheric composition changes, promoting global warming.

12

Population and Growth

12.1 INTRODUCTION

12.1.1 Definitions

Demography Study of human population dynamics is called demography. The study includes analysis of size, structure and distribution of human population of a particular region with respect to birth, migration, age and death. Population of a given region is divided into four major categories: (i) Nationality, (ii) Religion, (iii) Ethnicity and (iv) Education. In broader sense, demography analyses the relationships between economic, social, cultural and biological processes influencing a population.

Crude birth rate The number of live births per 1000 population per year is called crude birth rate of that year.

Total fertility rate The average number of children that would be born alive to a woman is called total fertility rate (TFR). It is always greater than or equal to 1. However, if TFR is expressed with respect to total population (including male and female), it is often a fractional quantity.

Replacement level fertility It refers to the minimum number of children per female such that at least one child is female category. Replacement level fertility cannot be achieved if all children are male category.

Crude death rate The number of deaths per 1000 population per year is called crude death rate of that year.

Infant mortality rate It refers to the number of deaths of infants (under one year of age) per 1000 live births in a given year. It is also called poverty index of a country. In the poor and underdeveloped countries infant mortality rate may be as high as 200 per 1000 per year but in the developed countries the infant mortality rate is very low, around 9 per 1000 per year.

12.1.2 Population momentum

Even after achievement of replacement level fertility population continues to grow since the number of births overshadows the number of deaths. This phenomenon of continuation of population growth after achieving replacement level fertility is known as population momentum.

Crude birth rate (*b*) is usually greater than crude death rate (*d*) and the difference (*r*), $r = (b - d)$, is called natural increase. At an early stage *r* is very high but birth rate decreases with increase in economic and social development and death rate also decreases due to the following reasons

(a) Better and assured food supply.
(b) Improved sanitation.
(c) Modern medicines.

Population of an area is always increasing due to migration of people. If *m* is the migration rate, the expression of *r* will be $r = (b - d + m)$. Thus, a transition from high birth and death rates to low birth and death rates is observed. This transition is called *demographic transition*. When birth rate and death rate are constant the population of any age category remains constant. This is called *stable population.*

12.2 LIFE TABLE

In a given location life table provides valuable data regarding (i) current age structure, (ii) mortality data and (iii) future population projection.

P_0 = Total population at the starting point considering all age categories.

L_0 = Number of children in the age categories 0–4 years who would be alive in the next five years.

P_s = Number of children in the age categories 5–9 years who would be alive in the next five years.

Similarly, P_{10}, P_{15} .... etc. are also defined.

b_x = Number of surviving children born per person in age category *x* to $(x + 5)$ years. It is assumed that woman's fertility begins at 15 years of age and ends at 30 years of age. It is also known as fertility.

b_{15} = Number of surviving children born per person in age category 15 to 20 years.

b_{20} = Number of surviving children born per person in age category 20 to 25 years.

b_{25} = Number of surviving children born per person in age category 25 to 30 years.

All other values of b_x are zero.

Assume that $b_{15} = 0.37$, $b_{20} = 0.40$, $b_{25} = 0.30$.

$\frac{L_{x+5}}{L_x}$ represents the probability of those people of age category *x* to be alive after 5 years.

Let us consider the following life table

Table 12.1 Life table for population projection.

Age category	$\frac{L_{x+5}}{L_x}$	*Fertility* b_x	*1985*	*1990*
0–4	0.971	0	18000	P_0
5–9	0.991	0	17000	P_1
10–14	0.991	0	16000	P_2
15–19	0.988	0.37	19000	P_3
20–24	0.986	0.40	20000	P_4
25–29	0.986	0.30	22000	P_5
30–34	0.983	0	20000	P_6
35–39	0.978	0	19500	P_7
40–44	0.978	0	19000	P_8
45–49	0.975	0	18500	P_9

$P_1 = 0.971 \times 18000 = 17478,$ $P_2 = 0.991 \times 17000 = 16847.$
$P_3 = 0.991 \times 16000 = 15856,$ $P_4 = 0.988 \times 19000 = 18772.$
$P_5 = 0.986 \times 20000 = 19720,$ $P_6 = 0.986 \times 22000 = 21692.$
$P_7 = 0.983 \times 20000 = 19660,$ $P_8 = 0.978 \times 19500 = 19071.$
$P_9 = 0.978 \times 19000 = 18582.$
$P_0 = 0.37 \times 19000 + 0.40 \times 20000 + 0.30 \times 22000 = 21630.$

Total population in 1985 = 189000.

Population projection in 1990 = $P_0 + P_1 + P_2 + P_3 + P_4 + P_5 + P_6 + P_7 + P_8 + P_9$
= 189308.

12.3 POPULATION GROWTH MODEL

Population growth of a region may be calculated by any one of the following methods

1. Discrete growth.
2. Exponential growth.
3. Logistic growth.

12.3.1 Discrete growth model

Population of a region is enumerated separately for several years and an average growth rate r is considered. Then we can write

$$N_1 = N_0(1 + r), \quad N_2 = N_1(1 + r) = N_0(1 + r)^2, \quad N_3 = N_2(1 + r) = N_0(1 + r)^3$$

$$N_t = N_{t-1}(1 + r) = N_1(1 + r)^t$$

$$N_t = N_0(1 + r)^t \tag{12.1}$$

12.3.2 Exponential growth model

Rate of increase of population is directly proportional to the existing number of population only. Proportionality constant is called average growth rate constant, r.

$$\frac{dN}{dt} \propto N.$$

Or,
$$\frac{dN}{dt} = rN \tag{12.2}$$

On integration we get $N_t = N_0 e^{rt}$ (12.3)

The exponential growth curve is shown in the Fig. 12.1 below.

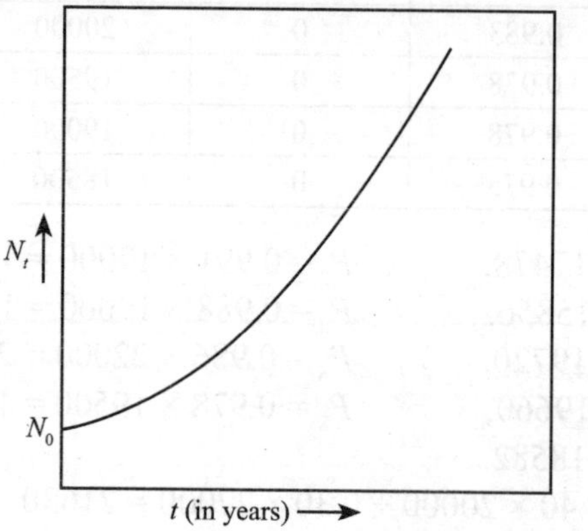

Fig. 12.1 Exponential growth curve.

Doubling time (t_d) The time required to double up the population is called doubling time, t_d. $N_t = 2N_0$. Putting this value of N_t in the above equation (12.3), we get:

$$t_d = \frac{0.693}{r}$$

Half-life period ($t_{1/2}$) The time required to decay the population to half of the original population is called half-life period $t_{1/2}$.

$$-\frac{dN}{dt} \propto N \qquad \text{Or,} \quad \frac{dN}{dt} = -rN$$

On integration we get

$$N_t = N_0 e^{-rt}. \qquad \text{Or,} \quad \ln\frac{N_0}{N_t} = rt \tag{12.3a}$$

At $t = t_{1/2}$, $N_t = \frac{1}{2}N_0$. Using this boundary condition in the above equation (12.3a), we get

$$t_{1/2} = \frac{0.693}{r}$$

12.3.3 Logistic growth model

According to this model continuous exponential growth of population cannot occur due to (i) limited food supply and (ii) environmental pollution. A study was conducted on a species of bacteria as shown in the Fig. 12.2 below. Different phases are marked as *a*, *b*, *c* and *d*.

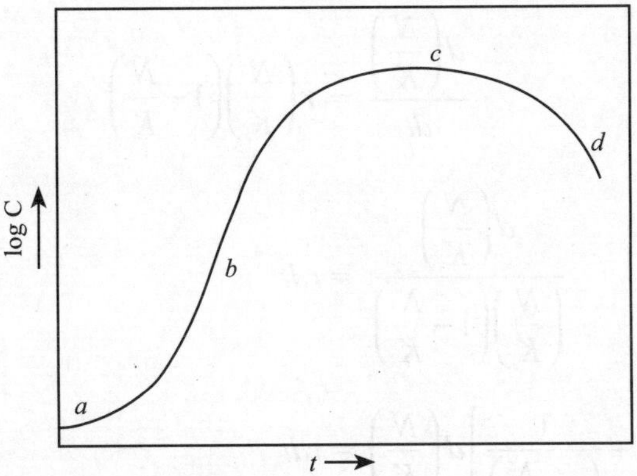

Fig. 12.2 Population growth curve of a species of bacteria.

Lag phase (*a*) In this phase population growth is minimum since organisms need some time period to be adapted with the new environment.

Exponential phase (*b*) Population growth is very rapid. Food supply is enough to grow population at a very fast rate.

Stationary phase (*c*) In this phase population reaches a maximum, food supply is limited and further growth ceases to occur. The maximum population is called *carrying capacity* of the environment. So carrying capacity is defined as the maximum population that the environment can carry without any adverse effect. It is represented by the symbol *K*.

Death phase (*d*) At this phase environment becomes hostile sufficiently so that the population starts declining. Environmental resistance is dominating at this phase.

Considering all the phases the population growth equation can be modified to the following form

$$\frac{dN}{dt} \propto N\left(1-\frac{N}{K}\right)$$

Or,
$$\frac{dN}{dt} = rN\left(1-\frac{N}{K}\right) \tag{12.4}$$

where, $r =$ average growth rate constant.

$K =$ Carrying capacity of the environment.

$\left(1-\frac{N}{K}\right) =$ Environmental resistance.

12.3.3.1 Environmental resistance

When population reaches carrying capacity of the environment, food supply is limited and the environment resists any further growth of population. This is called environmental resistance.

The equation (12.4) can be written as

Or, $$\frac{d\left(\frac{N}{K}\right)}{dt} = r\left(\frac{N}{K}\right)\left(1-\frac{N}{K}\right)$$

Or, $$\frac{d\left(\frac{N}{K}\right)}{\left(\frac{N}{K}\right)\left(1-\frac{N}{K}\right)} = rdt$$

Or, $$\left[\frac{1}{\left(\frac{N}{K}\right)}+\frac{1}{\left(1-\frac{N}{K}\right)}\right]d\left(\frac{N}{K}\right) = rdt$$

Or, $$\int\frac{d\left(\frac{N}{K}\right)}{\left(\frac{N}{K}\right)}+\int\frac{d\left(\frac{N}{K}\right)}{\left(1-\frac{N}{K}\right)} = \int rdt + C$$

Or, $$\ln\frac{\left(\frac{N}{K}\right)}{\left(1-\frac{N}{K}\right)} = rt + C$$

Or, At $t = \tau$, $\quad \mathrm{N} = \frac{K}{2}$

Using this boundary condition in the above equation, we get $C = -r\tau$

Or, $$\ln\frac{\left(\frac{N}{K}\right)}{\left(1-\frac{N}{K}\right)} = rt - r\tau = r(t-\tau)$$

Or, $$\ln\frac{\left(1-\frac{N}{K}\right)}{\left(\frac{N}{K}\right)} = -\,r(t-\tau)$$

Or, $$\ln\left(\frac{K}{N}-1\right) = -r(t-\tau) \tag{12.5}$$

Or, $$\frac{K}{N} - 1 = e^{-r(t-\tau)}.$$

Or, $$\frac{K}{N} = 1 + e^{-r(t-\tau)}$$

Or, $$N = \frac{K}{1+e^{-r(t-\tau)}} \tag{12.6}$$

Equations (12.5) and (12.6) are known as logistic equations of population growth. The logistic growth curve is shown in the Fig. 12.3 below.

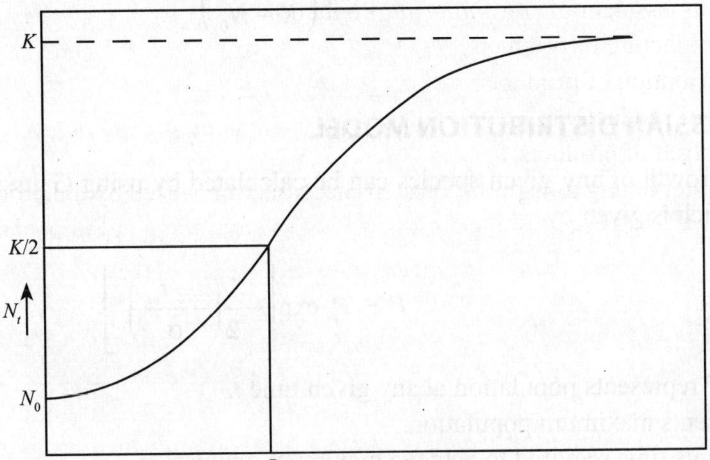

Fig. 12.3 Logistic growth curve.

At $t = 0$, $N = N_0$. Using this boundary condition in the equation (12.5), we get

$$\ln\left(\frac{K}{N_0}-1\right) = r\tau.$$

Or, $$\tau = \frac{1}{r}\ln\left(\frac{K}{N_0}-1\right) \tag{12.7}$$

12.3.3.2 Relation between *r* and R_0

It has been assumed that initially, i.e., around $N = N_0$, logistic growth rate is equal to exponential growth rate. If R_0 is the initial exponential growth rate constant, equation (12.2) becomes

$$\left|\left(\frac{dN}{dt}\right)\right|_{t\approx 0} = R_0 N_0$$

Again, the initial logistic growth rate can be calculated from equation (12.4)

$$\left|\left(\frac{dN}{dt}\right)\right|_{t\approx 0} = rN_0\left(1-\frac{N_0}{K}\right)$$

The above two equations lead to the following result

$$rN_0\left(1-\frac{N_0}{K}\right) = R_0 N_0$$

Or, $$r = R_0\left(\frac{K}{K-N_0}\right) \tag{12.8}$$

12.4 GAUSSIAN DISTRIBUTION MODEL

Population growth of any given species can be calculated by using Gaussian distribution function, which is given by

$$P = P_m \exp\left[-\frac{1}{2}\left(\frac{t-t_m}{\sigma}\right)^2\right] \tag{12.9}$$

Where, P represents population at any given time t.

P_m represents maximum population.

t_m represents time required to achieve maximum population.

σ is the variance.

The Gaussian distribution curve is shown in the Fig. 12.4. The Gaussian distribution function is widely used to estimate the total amount of a non-renewable resource like petrol, minerals etc., which are continuously consumed. If Q represents the total reserve of any consumable resource, we can write that

$$Q = \int_{-\infty}^{+\infty} P.dt,$$

where, P represents consumption rate.

$$Q = \int_{-\infty}^{+\infty} P.dt = 2P_m\int_{0}^{\infty} e^{-x}\left(\sigma\sqrt{2x}\right)dx$$

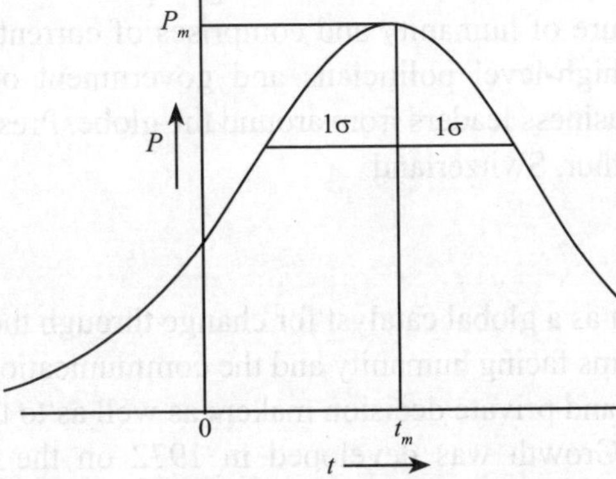

Fig. 12.4 Gaussian distribution curve.

$$\text{where,} \quad x = \frac{1}{2}\left(\frac{t-t_m}{\sigma}\right)^2 \quad \text{and hence,} \quad dt = \sigma\sqrt{2x}dx$$

$$Q = 2\sqrt{2}\sigma P_m\int_0^\infty \sqrt{x}e^{-x}dx = 2\sqrt{2}\sigma P_m\int_0^\infty x^{\frac{3}{2}-1}e^{-x}dx = 2\sqrt{2}\sigma P_m\Gamma\left(\frac{3}{2}\right) = 2\sqrt{2}\sigma P_m\Gamma\left(\frac{1}{2}+1\right)$$

$$Q = 2\sqrt{2}\sigma P_m\frac{1}{2}\Gamma\left(\frac{1}{2}\right) = 2\sqrt{2}\sigma P_m\frac{1}{2}\sqrt{\pi} = \sigma P_m\sqrt{2\pi}$$

To find out the value of t_m put $t = 0$ in the equation (12.9): At $t = 0$, $P = P_0$

$$\text{Hence,} \quad P_0 = P_m \exp\left[-\frac{1}{2}\left(\frac{-t_m}{\sigma}\right)^2\right]$$

$$\text{Hence,} \quad t_m = \sigma\sqrt{2\ln\frac{P_m}{P_0}}$$

12.5 CLUB OF ROME

12.5.1 History of foundation

The Club of Rome was founded in April 1968 at Accademia dei Lincei in Rome, Italy by Aurelio Peccei, an Italian industrialist, and Alexander King, a Scottish scientist. It was formed when a small international group of people from the fields of academia, civil society, diplomacy, and industry, met at a villa in Rome, Italy, hence the name.

The Club of Rome (CoR) is a global think tank, dealing with different political and apolitical issues regarding environment. The CoR is a group of world citizens, sharing a common concern for the future of humanity and comprises of current and former Heads of State, UN bureaucrats, high-level politicians and government officials, diplomats, scientists, economists, and business leaders from around the globe. Present headquarters of the organization is in Winterthur, Switzerland.

12.5.2 Mission of CoR

The mission of CoR is "to act as a global catalyst for change through the identification and analysis of the crucial problems facing humanity and the communication of such problems to the most important public and private decision makers as well as to the general public".

The concept Limits to Growth was developed in 1972 on the basis of computer modeling of economic and population growth with finite resource supplies. Exponential growth model was taken as a standard to develop computer modeling. There are five important variables

(i) World population.
(ii) Industrialization.
(iii) Pollution.
(iv) Food production.
(v) Resource depletion.

12.6 RESERVE INDEX

Reserve index of minerals and oils can be expressed either by (i) static reserve index number or by (ii) exponential reserve index number.

12.6.1 Static reserve index number

To calculate the static reserve index number of a mineral/oil it is assumed that usage of that resource is constant. For example, in 1972, the amount of chromium reserves was 775 million metric tons and the fixed annual mining of chromium is 2 million metric tons. So

$$\text{Static reserve index number} = \frac{775}{2} = 387.5 \text{ years} \approx 388 \text{ years}$$

12.6.2 Exponential reserve index number

Let us consider Q be the total reserve of a resource and P is the production rate of the resource at any time t. P_0 is the initial production rate and the production rate increases exponentially as $P = P_0e^{rt}$. Thus we can write that

$$\frac{dQ}{dt} = P \quad \text{and} \quad P = P_0e^{rt}$$

'r' is the production rate constant.

$$\frac{dQ}{dt} = P_0 e^{rt}.$$

$$\int dQ = \int_0^T P_0 e^{rt} dt$$

Or,
$$Q = \frac{P_0}{r}\left[e^{rt}\right]_0^T = \frac{P_0}{r}\left(e^{rT} - 1\right)$$

Or,
$$T = \frac{1}{r}\ln\left[1 + \frac{Qr}{P_0}\right] \quad (12.10)$$

T is the total time required to consume the total resource and also known as *exponential reserve index number*.

12.7 MAXIMUM SUSTAINABLE YIELD (MSY)

12.7.1 Definition

Population of any species increases due to increase in growth rate, survival rate and reproductive rate. On the other hand population decreases due to increase in death rate and harvesting rate. So the population growth usually follows logistic growth model as shown in the Fig. 12.3. The logistic growth equation is

$$\frac{dN}{dt} = rN\left(1 - \frac{N}{K}\right) \quad (12.11)$$

The corresponding plot of growth rate (dN/dt) versus population size (N) is shown in the Fig. 12.5.

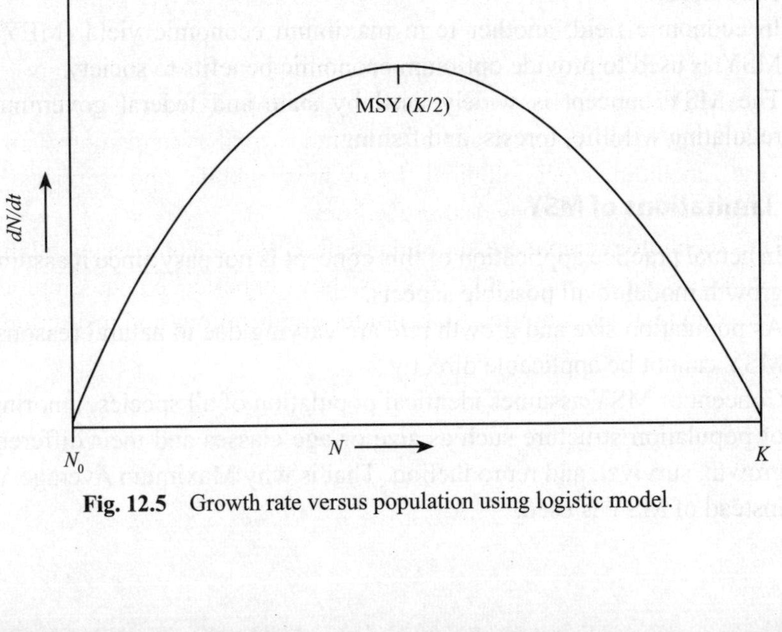

Fig. 12.5 Growth rate versus population using logistic model.

dN/dt indicates growth rate or sustainable yield. The curve shows that sustainable yield becomes maximum at a certain population, at which sustainable yield is called maximum sustainable yield or MSY. To find out MSY 1st derivative of *dN/dt* is set to zero.

$$\frac{d^2N}{dt^2} = r\left[\left(1-\frac{N}{K}\right)\frac{dN}{dt} - \frac{N}{K}\frac{dN}{dt}\right] = 0.$$

Or, $$N = \frac{K}{2}$$

Putting this value of N in equation (12.11), we get

$$MSY = \left|\frac{dN}{dt}\right|_{\max} = \frac{rK}{4} = \frac{R_0 K}{4\left(1-\frac{N_0}{K}\right)} \tag{12.12}$$

12.7.2 Significances of MSY

(a) In a given ecosystem, if the number of individuals, greater than its MSY, is harvested, the population will decline to extinction.
(b) If harvesting is carried out at the MSY level, population size of that species will remain constant.
(c) Harvesting below MSY level leads to a stable equilibrium population.

12.7.3 Applications of MSY

(a) The term MSY is widely used in fisheries. For a given pond ecosystem it represents the maximum possible harvesting of fishes without disturbing the original population size.
(b) It helps us give an idea about optimum use of stock under existing environmental conditions.
(c) In economic field, another term maximum economic yield (MEY), similar to MSY, is used to provide optimum economic benefits to society.
(d) The MSY concept is widely used by state and federal government agencies regulating wildlife, forests, and fishing.

12.7.4 Limitations of MSY

(a) In actual practice application of this concept is not easy since it assumes a definite growth model to all possible aspects.
(b) As population size and growth rate are varying due to natural reasons, concept of MSY cannot be applicable directly.
(c) Concept of MSY assumes identical population of all species, ignoring all aspects of population structure such as size or age classes and their differential rates of growth, survival, and reproduction. That is why Maximum Average Yield (MAY) instead of MSY is used.

12.8 SOLVED PROBLEMS AND EXERCISES

Problem 1 Consider the following age structure with 3 categories

0–24 years	2.5 billion
25–49 years	1.5 billion
50–74 years	1.0 billion

Assume that (i) all births occur on the woman's 25th birthday. (ii) All deaths occur at the age of 75. (iii) The male to female ratio is always 1:1.

The total fertility rate (TFR) is 4 for the 1st 25 years and then it drops instantaneously to 2. Draw the age structures after 25, 50 and 75 years separately.

Solution The initial age structure, i.e., at $t = 0$, is shown below

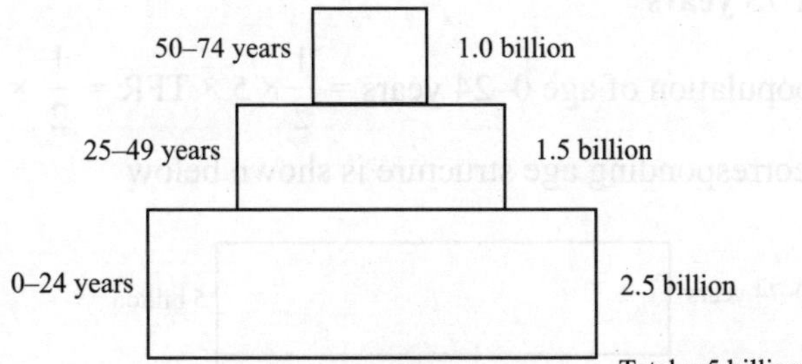

Initial age structure at $t = 0$

As TFR is related to only female category, women population, i.e., half of the total population, is considered to calculate the total number of births.

After 25 years

$$\text{Net population of age 0–24 years} = \frac{1}{2} \times 2.5 \times \text{TFR} = \frac{1}{2} \times 2.5 \times 4 = 5 \text{ billion.}$$

The corresponding age structure is shown below

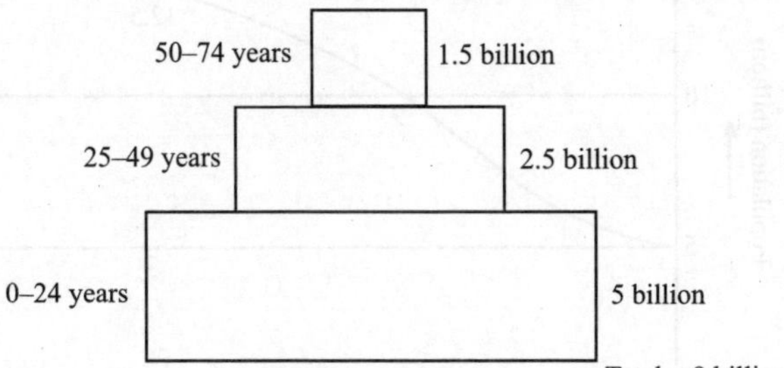

Age structure after 25 years

After 50 years

Net population of age 0–24 years = $\frac{1}{2} \times 5 \times \text{TFR} = \frac{1}{2} \times 5 \times 2 = 5$ billion.

The corresponding age structure is shown below

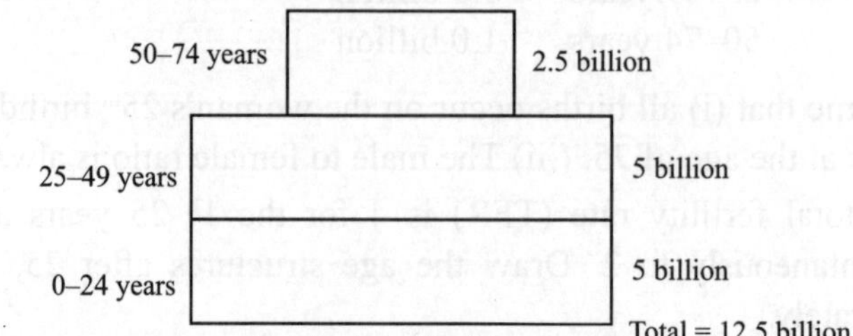

Age structure after 50 years

After 75 years

Net population of age 0–24 years = $\frac{1}{2} \times 5 \times \text{TFR} = \frac{1}{2} \times 5 \times 2 = 5$ billion.

The corresponding age structure is shown below

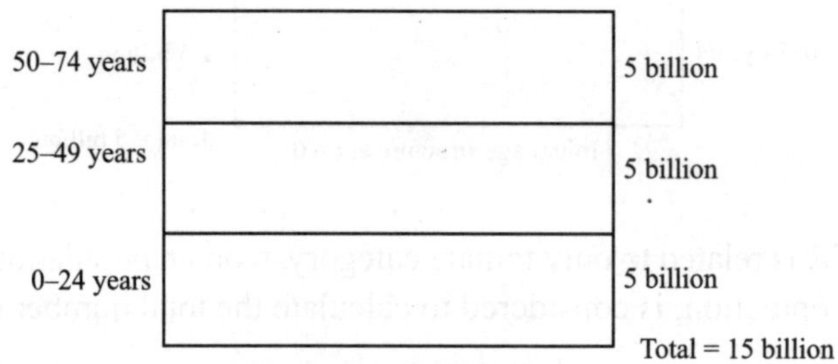

Age structure after 75 years

The population curve is shown below

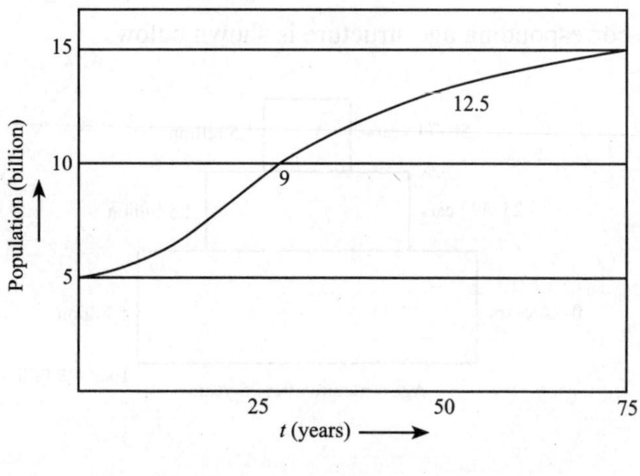

After 25 years population of age 0–24 years is equal to the population of age 25–49 years after 50 years. So replacement level fertility is achieved after 25 years. The above figure shows that population is increasing continuously even after reaching replacement level fertility. The phenomenon is called population momentum.

Problem 2 In a given area an age structure has three categories

Age range	*Population*
0–24 years	3 billion
25–49 years	2 billion
50–74 years	1 billion

Assume the following conditions

(i) All births occur on the woman's 25th birthday.
(ii) All deaths occur after completion of 74 years.
(iii) Replacement level fertility exists from the starting point.
(iv) Male:female ratio is assumed to be 1:1 in all categories of population.
(v) The TFR is constant and its value is 2.

Draw the age structures after 25, 50 and 75 years, indicating the population size.

Solution The initial age structure is shown in the following figure

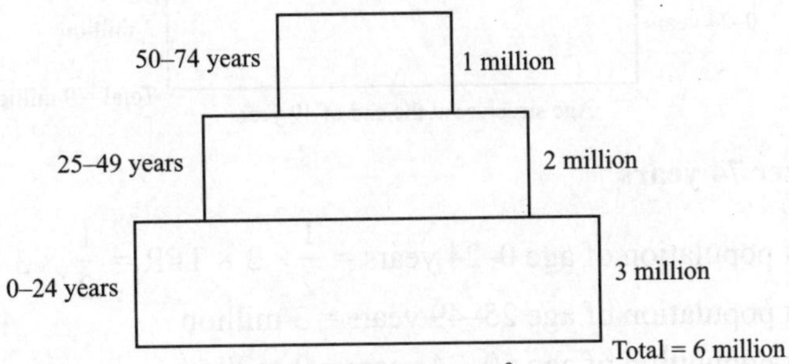

Initial age structure at $t = 0$

After 24 years

Net population of age 0–24 years $= \frac{1}{2} \times 3 \times \text{TFR} = \frac{1}{2} \times 3 \times 2 = 3$ million.

Net population of age 25–49 years = 3 million.

Net population of age 50–74 years = 2 million.

The corresponding age structure is shown below

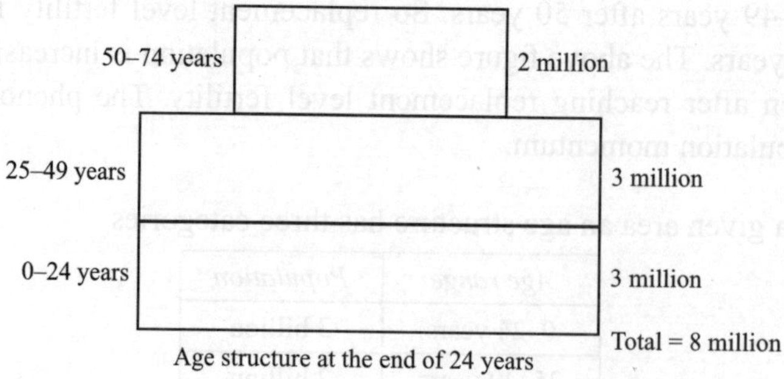

Age structure at the end of 24 years

After 49 years

Net population of age 0–24 years $= \frac{1}{2} \times 3 \times \text{TFR} = \frac{1}{2} \times 3 \times 2 = 3$ million.

Net population of age 25–49 years = 3 million.

Net population of age 50–74 years = 3 million.

The corresponding age structure is shown below

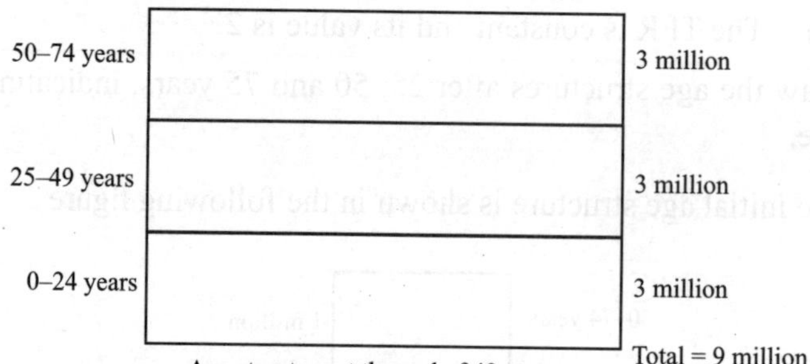

Age structure at the end of 49 years

After 74 years

Net population of age 0–24 years $= \frac{1}{2} \times 3 \times \text{TFR} = \frac{1}{2} \times 3 \times 2 = 3$ million.

Net population of age 25–49 years = 3 million.

Net population of age 50–74 years = 3 million.

The corresponding age structure is shown below

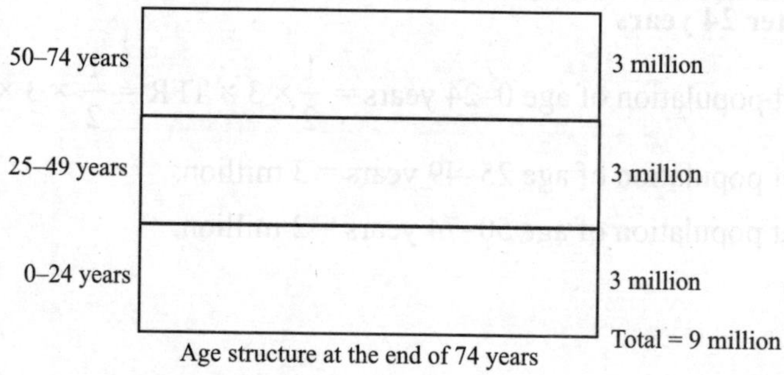

Age structure at the end of 74 years

Problem 3 In a given area age structure of human population has three categories

Age range	*Population*
0–19 years	2×10^5
20–39 years	8×10^4
40–59 years	1×10^4

Assume that the following conditions hold good

(i) All births occur on the woman's 20th birthday.
(ii) All deaths occur on the 60th birthday.
(iii) Replacement level fertility exists from the starting point.
(iv) Male:female ratio is assumed to be 1:1 in all categories of population.
(v) Population is growing at a constant rate of 3.5% per year.

(a) Calculate the TFR.
(b) Draw the age structures after 20, 40 and 60 years, indicating the population size.

Solution Assume that TFR is x.

After 20 years

Net population of age 0–19 years $= \frac{1}{2} \times 0.2 \times \text{TFR} = \frac{1}{2} \times 0.2 \times x$

$= 0.1x$ million.

Net population of age 20–39 years = 0.2 million.

Net population of age 40–59 years = 0.08 million.

Total population after 20 years $= N_{20}$.

$$N_{20} = (0.1x + 0.2 + 0.08) \text{ million} = (0.28 + 0.1x) \text{ million} \quad \text{(P 3.1)}$$

Using discrete model total population after 20 years $= N_{20}$, where, $r = 0.035$.

$N_{20} = N_0(1 + r)^{20}$, where, $N_0 = 0.29$ million and $r = 0.035$.

Hence,

$$N_{20} = 0.29 \times (1 + 0.035)^{20} = 0.29 \times (1.035)^{20} = 0.577 \text{ million} \quad \text{(P 3.2)}$$

Comparing equations (P 3.1) and (P 3.2), we have

$$0.28 + 0.1x = 0.577.$$

Or, $0.1x = 0.297$.

Or, $x = 2.97 \approx 3$.

Thus TFR is 3.

The initial age structure is shown below

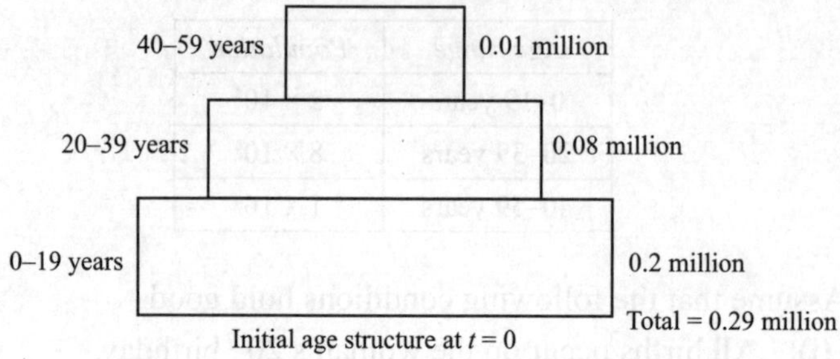

Initial age structure at $t = 0$

After 20 years

$$\text{Net population of age 0–19 years} = \frac{1}{2} \times 0.2 \times \text{TFR} = \frac{1}{2} \times 0.2 \times 3$$
$$= 0.3 \text{ million.}$$

Net population of age 20–39 years = 0.2 million.
Net population of age 40–59 years = 0.08 million.

The corresponding age structure is shown below

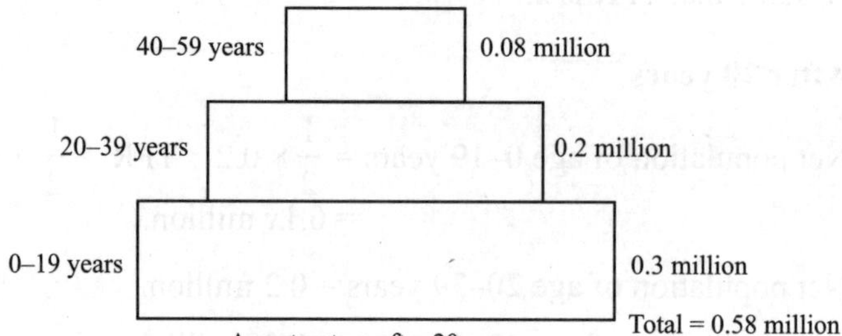

Age structure after 20 years

After 40 years

$$\text{Net population of age 0–19 years} = \frac{1}{2} \times 0.3 \times \text{TFR} = \frac{1}{2} \times 0.3 \times 3$$
$$= 0.45 \text{ million.}$$

Net population of age 20–39 years = 0.3 million.
Net population of age 40–59 years = 0.2 million.

The corresponding age structure is shown below

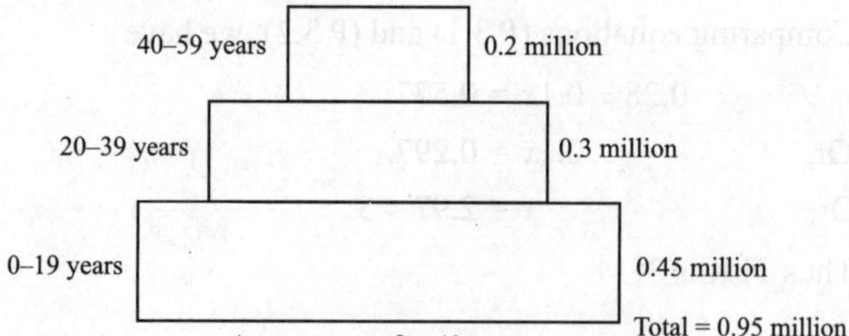

Age structure after 40 years

After 60 years

Net population of age 0–19 years $= \frac{1}{2} \times 0.45 \times \text{TFR} = \frac{1}{2} \times 0.45 \times 3$
$= 0.675$ million.

Net population of age 20–39 years = 0.45 million.
Net population of age 40–59 years = 0.3 million.
The corresponding age structure is shown below

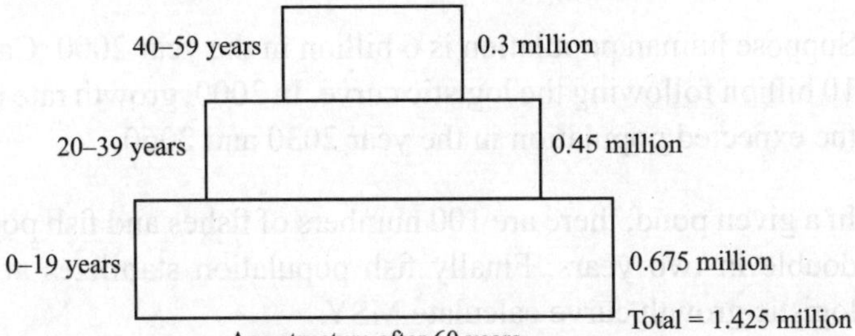

Age structure after 60 years

Problem 4 In 1960, the world's population was 3 billion and growth rate was 1.5% per year. Steady-state population is 10 billion. When would the population reach 5 billion? What would be the projected population in 1990 and 2015 using logistic model?

Solution

$$r = \frac{R_0}{1 - \frac{N_0}{K}}, \qquad \tau = \frac{1}{r}\ln\left(\frac{K}{N_0} - 1\right), \qquad t = \tau - \frac{1}{r}\ln\left(\frac{K}{N} - 1\right)$$

K = Carrying capacity = 10 billion.
R_0 = Initial growth rate = 0.015.
N_0 = Initial population = 3 billion.
τ = Time required to reach the population $K/2$ = 5 billion.

$$r = \frac{0.015}{1 - \frac{3}{10}}, \qquad \tau = \frac{1}{0.0214}\ln\left(\frac{10}{3} - 1\right) = 39.59 \text{ years.}$$

$$t_{1990} = 1990 - 1960 = 30 \text{ years} = 39.59 - \frac{1}{0.0214}\ln\left(\frac{10}{N} - 1\right)$$

$$\ln\left(\frac{10}{N} - 1\right) = (39.59 - 30) \times 0.0214 = 0.205. \quad N = 4.49 \text{ billion.}$$

$$t_{2015} = 2015 - 1960 = 55 \text{ years} = 39.59 - \frac{1}{0.0214}\ln\left(\frac{10}{N} - 1\right)$$

$$\ln\left(\frac{10}{N} - 1\right) = (39.59 - 55) \times 0.0214 = -0.33. \quad N = 5.82 \text{ billion.}$$

Exercise 1 World population in 1850 has been estimated to have been about 1 billion. It reached 4 billion in 1975. Estimate (i) the growth rate using discrete model and exponential model, (ii) time required to reach the population 2 billion using exponential model.

Exercise 2 A lake has carrying capacity of 20,000 fish. Current rate of fishing is 2000 fish per year and the fish population is assumed to be steady at about 6000. Calculate MSY.

Exercise 3 Suppose human population is 6 billion in the year 2000. Carrying capacity is 10 billion following the logistic curve. In 2000, growth rate is 1.3%. Calculate the expected population in the year 2030 and 2060.

Exercise 4 In a given pond, there are 100 numbers of fishes and fish population becomes double in two years. Finally fish population stabilizes at 5000. Assuming logistic growth curve calculate MSY.

13

Sustainable Development

13.1 ENVIRONMENTAL PROBLEMS

13.1.1 Some facts

(a) Around 1×10^9 people are using insalubrious water. One third of world's population has inadequate sanitation.
(b) 1.3×10^9 people are suffering from smoke and shoot problems.
(c) $(300 - 700) \times 10^6$ people are suffering from indoor air pollution.
(d) Farmers, forest dwellers and tribes are suffering from deforestation and degradation of land.
(e) Fast increase in world population is a threat. Sufficient agricultural and industrial developments are required to combat the population explosion.

13.1.2 Environmental problems in India

There are several environmental problems in India. These are listed below
(a) Water pollution and water scarcity.
(b) Air pollution.
(c) Soil pollution.
(d) Solid and hazardous wastes.
(e) Deforestation.
(f) Loss of biodiversity.
(g) Atmospheric changes.
(h) Population growth and poverty.

13.2 SUSTAINABLE DEVELOPMENT

Development is a comprehensive, economic, social, cultural and political process aiming at the constant improvement of all individuals on the basis of their active, free and meaningful participation in development. Thus sustainable development is that development which can protect the environment by satisfying the needs of the present generation keeping in view of the needs for development of future generation.

13.2.1 Necessity of sustainable development

Economic and technological developments sufficiently degrade the environment in different areas, e.g.,

(a) Depletion of local resource base and misuse of resources.
(b) Depletion of ozone layer.
(c) Degradation of soil quality and deforestation which causes flooding at low land areas.
(d) Direct discharge of domestic and industrial effluents to water reservoirs threatens the aquatic life.
(e) Continuous acid rain in industrial areas and nuclear fall out across the borders.
(f) Use of hazardous chemicals and chemical weapons.
(g) Resource gap between developing and developed countries.
(h) Global warming and consequent change in climate has pronounced effect on biodiversity, plant and living organisms including human being.

13.2.2 Some examples of environmental degradation for the sake of economic development

(a) Construction of dams at Tehri Garhwal and Navagam in Gujarat resulted in a huge loss of agricultural land. Many people have lost their homes and they are called *"environmental refugees"*. This is an example of an economic development, which is contrary to sustain environment.
(b) In Thailand and Malaysia several acres of agricultural land are converted to industrial belt, as a result of which, thousands of villagers became environmental refugees.
(c) An international chemical company, "HOECHST" has several factories in the world. The main product of the company is insecticides. During the manufacturing of insecticides poisonous gases are emitted, causing severe health hazards.

The existence of life on Earth is threatened due to rapid economic and technological developments. So people should concentrate on those developments which are eco-friendly. Thus, sustainable development is very much necessary for survival of life on Earth.

13.2.3 Concept of sustainable development

It considers the following points

(a) Complete elimination of poverty and deprivation is essential to establish sustainable development.
(b) Conservation and enhancement of resource base is another important step in support of sustainable development.

(c) Sustainable development does not mean the economic growth only. It should consider social and cultural development also.

(d) Sustainable development always considers unification of ecology and economics at all levels.

13.2.4 Steps to execute the above concept of sustainable development

(a) Reciprocal rights and developments must be recognised with respect.

(b) Some common minimum programmes must be established to ease tension and improve interstate behaviour.

(c) To abide by international agreements in support of sustainable development.

(d) Some procedures must be adopted to avoid all existing environmental disputes.

13.3 AGENDA 21

Countries of northern hemisphere first made an attempt to establish an international association in 1990 in support of sustainable development. It is known as International Council for Local Environmental Initiatives or ICLEI. The first conference was convened in September, 1990 at the United Nations in New York, where 200 local governments from 43 countries took part. It provides a basis of international performance, result-oriented campaigns and programs.

13.3.1 Objectives of ICLEI

(i) It provides technical consulting, training and information services to develop capacity and infrastructure, share knowledge and also help local government implement sustainable development at the local level.

(ii) It helps companies and government to design efficient and cost-effective way to achieve local, national and global sustainability objectives.

(iii) It promotes Agenda 21 globally.

13.3.2 The agreement

The 1st international conference on sustainable development was held in 1992, at Rio de Janerio, where leaders of developed and developing countries assembled. They agreed that current developmental practice was dangerous and would put our future generations in a tight spot. The conference passed Agenda 21 treaty, which enjoins both nations and local authorities to produce a sustainable development strategy.

The activities of companies have profound effect on the environment in which citizens live and people are entitled to be informed about such activities so as to be able to make fully informed decisions relating to their environment. According to Gray et al (1993): "Information and decision-making must be democratic in the widest sense of the term because it is the society as a whole which must have the right to choose the type of trades that are essential in support of sustainable development." Agenda 21 is the primary policy document emerged from *'Rio Earth Summit'* and identifies the right to know as a priority issue and calls on the international community, countries and individual companies to adopt it as a matter of environmental policy.

Agenda 21 is a 300-page document, divided into 40 chapters, grouped into 4 sections.

13.3.2.1 Section I: Social and Economic Dimensions

This section describes how to eliminate poverty, especially in developing countries, change consumption patterns, promote health, achieve stabilization of population and sustainable settlement in decision making.

13.3.2.2 Section II: Conservation and Management of Resources for Development

Includes atmospheric protection, combating deforestation, protecting fragile environments, conservation of biodiversity, control of pollution and the management of biotechnology and radioactive wastes.

13.3.2.3 Section III: Strengthening the Role of Major Groups

Includes the roles of children, youth, women, NGOs, local authorities, business and workers and strengthening the role of nations, their communities and farmers.

13.3.2.4 Section IV: Means of Implementation

Implementation includes science, technology transfer, education, international institutions and financial mechanisms.

13.4 WATER USE SUSTAINABILITY

That proper use and control of water resources helps our society develop and flourish into an indefinite future without degrading/destroying/disturbing the various components of the hydrologic cycle is called sustainable water use.

13.4.1 Criteria for water use sustainability

(a) Sufficient water supply to all sectors of an ecosystem is the prime criteria of water use sustainability.
(b) To discover new water resources to meet the huge water requirement.
(c) Minimum drinking water quality standard, as prescribed by WHO, should be maintained at all levels.
(d) To restrict those human activities which damage different water resources.
(e) Direct discharge of industrial and domestic effluents to the water resources must be prohibited.
(f) Waste water should be recycled after proper treatment.
(g) Action should be taken to minimize evaporation of water from surface water sources.

(h) Laws and procedures should be established to preserve ponds and other water resources.

13.5 RESOURCES

13.5.1 Definition

Resource is anything that we get from our environment to meet our needs and desires.

Necessity of resources Resources are essential to all human beings due to the following reasons

- Sustenance of life on Earth.
- Materials.
- Energy.

The natural resource base is limited and is insufficient to meet the demand of the tremendous growth of population. Thus proper planning is necessary for utilisation and conservation of resources.

13.5.2 Classification

Resources are of two types

1. **Material resource** Material resource can be divided into two categories
   (a) *Renewable resources* The sources of energy that are not depleted or exhausted by consumption and can be used continuously, are known as renewable resources or energy. These resources are continuously replenished through natural cycles. Examples are: solar energy, wind energy, tidal energy, biomass energy etc.
   (b) *Non-renewable resources* The energy sources, which are replenished slowly and cannot be replenished through natural cycles and hence may be exhausted on continuous using, are called non-renewable energy resources. Examples are: oils, fossil fuels, minerals etc.
2. **Intellectual resource** This basically includes intelligence, imagination, and capacity for work etc.

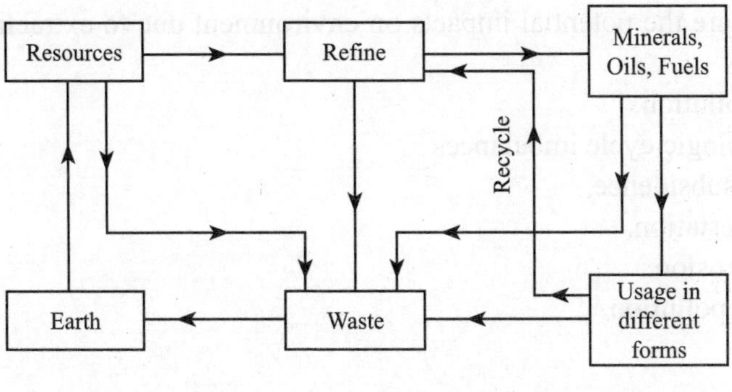

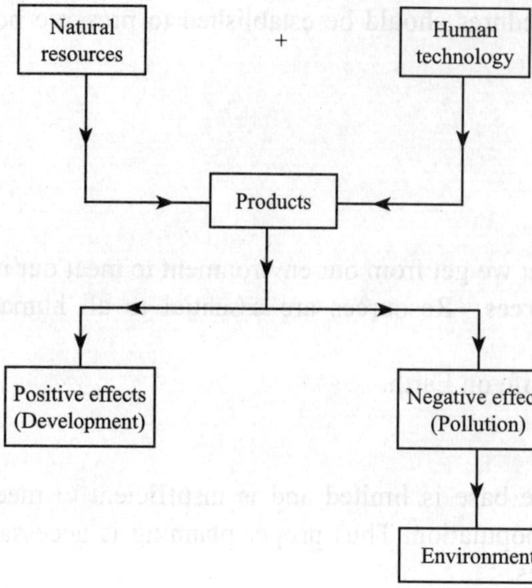

13.5.3 Mineral resource

- It is a non-renewable resource.
- Mining cost.
- Cost of discovery of new resource base.
- Increase in cost during advancement of production.

13.5.3.1 Global scenario

1. Coal, oil, minerals (Sb, Cr, Mn, Pb) are falling short of 20% in the near future.
2. Cu, Ag, Au, Ni, Sn are falling short of 80% by 2050 A.D.

13.5.3.2 Steps for conservation

(a) Improvement in extraction technology so that minerals can be extracted with minimum loss.
(b) Use of low grade minerals. Technology should be improved accordingly.
(c) Recovery, reuse and recycle of minerals.

Following are the potential impacts on environment due to extraction of coal, oil and minerals

(a) Air pollution.
(b) Hydrologic cycle imbalances.
(c) Land subsidence.
(d) Deforestation.
(e) Soil erosion.
(f) Noise pollution.

13.5.3.3 EIA suggestions

1. Comprehensive plan including the environmental report must be submitted before EIA authorities.
2. Baseline condition of the environment must be described.
3. Purpose and need of the project.
4. The future plan of the site must be projected properly.
5. Short term and long term impacts of the project should be assessed.

13.5.4 Biomass energy

Energy received from plant is called biomass energy. The total amount of organic matter in a particular ecosystem is called biomass. This organic matter can be burnt as solid fuel or converted into liquid or gaseous biofuels. The burning of wood and manure produces about 12% of the world's total energy. 28% of this biomass energy is used in the less developed countries in Asia and Africa.

13.5.5 Geothermal energy

It is the natural heat of the Earth. It is generated by the action of molten matter within the Earth and also from the radioactive decay of rocks. The main source of geothermal heat is hot water or dry stream. Hot springs or geysers are found in several places in India and world. Most of these hot water resources contain objectionable mineral matters that are polluting both the surface and subsurface water. Various gases (H_2S, SO_2 etc.) and radioisotopes are emitting from the geothermal sources and polluting the environment. During extraction of the geothermal sources ground subsidence appears as a major hazard.

13.6 RESOURCE CONSUMPTION

13.6.1 Exponential consumption

Let us consider Q be the total reserve of a resource and P is the production rate of the resource at any time t. P_0 is the initial production rate and the production rate increases exponentially as $P = P_0 e^{rt}$. Thus we can write that

$$\frac{dQ}{dt} = P. \quad \text{and} \quad P = P_0 e^{rt},$$

where, 'r' is the production rate constant.

$$\frac{dQ}{dt} = P_0 e^{rt}. \quad \int dQ = \int_0^T P_0 e^{rt}\, dt$$

Or,

$$Q = \frac{P_0}{r}\left[e^{rt}\right]_0^T = \frac{P_0}{r}\left(e^{rT} - 1\right)$$

Or, $$T = \frac{1}{r}\ln\left[1+\frac{Q\times r}{P_0}\right]$$

T is the total time required to consume the total resource.

13.6.2 Gaussian distribution of resource consumption

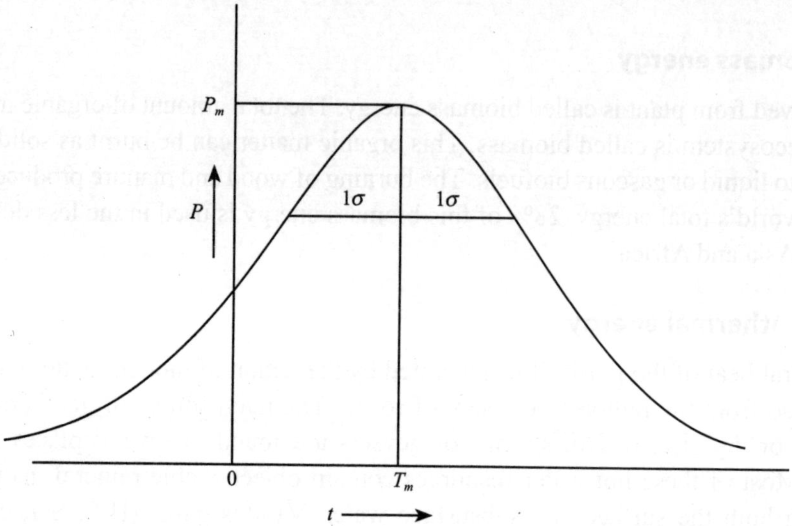

$$P = P_m \exp\left[-\frac{1}{2}\left(\frac{t-t_m}{\sigma}\right)^2\right] \tag{13.1}$$

$$Q = \int_{-\infty}^{+\infty} P.dt = 2P_m\int_0^{\infty} e^{-x}\left(\sigma\sqrt{2x}\right)dx$$

where, $x = \frac{1}{2}\left(\frac{t-t_m}{\sigma}\right)^2$ and hence, $dt = \sigma\sqrt{2x}dx$

$$Q = 2\sqrt{2}\sigma P_m\int_0^{\infty}\sqrt{x}e^{-x}dx = 2\sqrt{2}\sigma P_m\int_0^{\infty}x^{\frac{3}{2}-1}e^{-x}dx = 2\sqrt{2}\sigma P_m\Gamma\left(\frac{3}{2}\right) = 2\sqrt{2}\sigma P_m\Gamma\left(\frac{1}{2}+1\right)$$

$$Q = 2\sqrt{2}\sigma P_m\frac{1}{2}\Gamma\left(\frac{1}{2}\right) = 2\sqrt{2}\sigma P_m\frac{1}{2}\sqrt{\pi} = \sigma P_m\sqrt{2\pi}$$

To find out the value of t_m put $t = 0$ in the equation (13.1): At $t = 0$, $P = P_0$

Hence, $P_0 = P_m \exp\left[-\frac{1}{2}\left(\frac{t_m}{\sigma}\right)^2\right]$. Hence, $t_m = \sigma\sqrt{2\ln\frac{P_m}{P_0}}$

14

Environmental Management System

14.1 THE DEMING CYCLE

Dr. W. Edwards Deming, a pioneer in the field of quality management, proposed the following management cycle, known as Deming cycle.

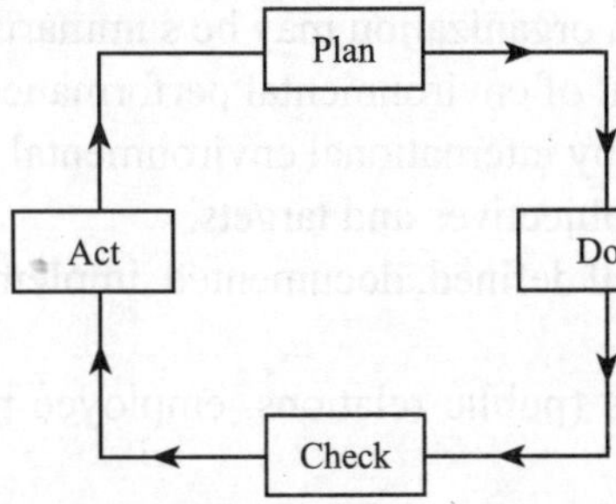

Fig. 14.1 The Deming cycle illustrating the management.

14.2 INTRODUCTION

Companies are committed to preserve the environment in their local communities and around the world by developing and implementing their EMS. ISO 14001 is a standard stating the requirements for an organization to develop and implement its EMS. It is used to establish and assess the effectiveness of an organization's ability to develop and meet a defined policy and objectives related to the environment.

14.3 OBJECTIVES OF EMS

(a) To develop company's environmental policy statement.
(b) To implement effectively and efficiently the programmes of environmental and conservation measures.
(c) To achieve associated objectives and targets by eliminating significant adverse effects, created due to organization's activities, on environment.

14.4 ISO 14001 ELEMENTS

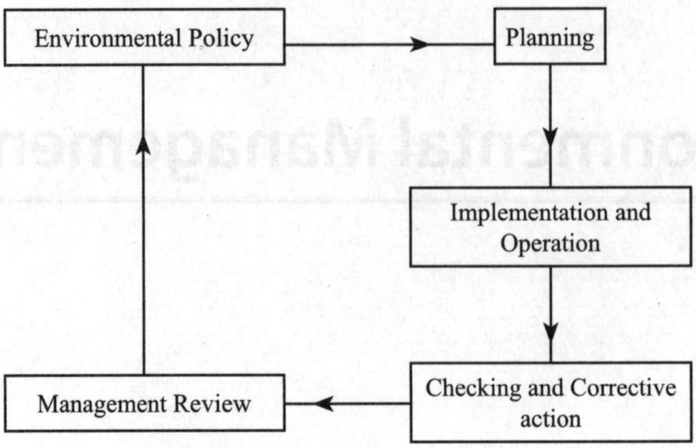

Fig. 14.2 ISO 14001 key elements.

14.4.1 Environmental policy

The environmental policy of an organization may be summarised as follows

1. Continuous assessment of environmental performance.
2. Commitment to abide by international environmental legislations and regulations.
3. To set environmental objectives and targets.
4. The policy must be well defined, documented, implemented and communicated to all employees.
5. Environmental aspects (public relations, employee participation, environmental performance etc.).
6. The policy must be made public.

14.4.2 Implementation and operation

1. **Infrastructure and responsibility** The organization must have sufficient resources to implement EMS. The resources include specialised skills, technology, material, energy and financial resources.
2. **Training awareness** It is required that each and every employee should be sufficiently aware of importance of environment, organization's impact on environment and also the benefits of implementation of EMS.
3. **Communication** The organization should have good communication system so that employees at all levels can communicate themselves smoothly. Besides that organization should have good global communication.
4. **EMS documentation and control** EMS documentation should be maintained and located properly. The documents should be regularly reviewed, revised as necessary and subsequently approved by authorised personnel.
5. **Emergency preparedness and response** The organization must be ready to cope with any emergency situation, related to environment.

14.4.3 Checking and corrective action

(a) Monitoring and measurement of company's environmental performance.
(b) Non-conformance corrective and preventive action for abatement of environmental pollution.
(c) Maintain records of company's environmental aspects.
(d) To execute EMS audit on a regular basis.

14.5 ISO 14001 REGISTRATION PROCESS

(a) Application/Contract

(i) Determination of scope of registration.
(ii) Location of organization.
(iii) Number of employees, amenities and facilities provided to them.

(b) Pre-audit option

Optional pre-audit can be used as an additional tool to determine the readiness of company's EMS.

(c) Technical review

This review must be conducted on-site. Without this visit a registration cannot be issued. This visit must be conducted at least six weeks before the registration audit. The following information is to be reviewed and evaluated

(i) Details of environmental policy of the organization.
(ii) Scope of implementing EMS in the organization.
(iii) Organization's adverse impacts on the environment and its determination of aspects to control environmental pollution.
(iv) Documentation of EMS, e.g., (i) machineries and equipment to purify solid, liquid and gaseous wastes of the plant, (ii) manuals of those machineries and equipment, (iii) proper documentation of procedures to execute those machineries and equipment, (iv) record of data obtained from those machineries and equipment and (v) proper maintenance of machineries and equipment.
(v) An overview of statutory regulations and permits regarding environmental pollution.

(d) Contents of environmental reports

(i) Organization's activities including production process, raw materials resource and energy resource, products, manpower etc.
(ii) Organization's EMS programme and its effectiveness to abate environmental pollution.

(iii) Organization's environmental performance with regard to its environmental objectives and targets.
(iv) Details of complaints and legal proceedings regarding environmental pollution.

(e) Documentation review

TUV Rhineland of North America Inc. will evaluate the readiness of organization for a formal audit. Only after review of the pre-audit report does the ISO 14001 authority allows the organization to go for a formal audit.

(f) Certification assessment

Technical review and Registration audit are to be conducted first. An on-site assessment will be conducted to confirm whether the organization's EMS complies with ISO 14001 standard requirements. Deviations are noted and the organization is asked to take necessary corrective action. After successful completion and review of the corrective action proposals, company will be recommended for registration.

(g) Registration audit

Thorough review of effectiveness of EMS is to be performed with respect to ISO 14001 and their internal procedures. Upon completion of the audit, an audit report is presented, describing audit results. The report also includes the deviations found with respect to ISO 14001. If the deviations are minor, registration audit will receive recommendation for registration by the audit team.

(h) Certification

After an independent review, TUV Rheinland will issue ISO 14001 certificate.

(i) Certification Maintenance (Surveillance)

Surveillance team visits to the organization annually to maintain the integrity of EMS.

14.5.1 Opportunities

Implementation of ISO 14001 at all levels of an organization can provide the following opportunities

(a) Reduced operating cost by implementing waste reduction, recycle and liability management.
(b) Increased opportunities for business in the global marketplace.
(c) Ability to meet public environmental expectations.

14.5.2 Benefits

The following benefits can be availed through implementation of ISO 14001 standard EMS

(a) Customer satisfaction.
(b) Employee participation in EMS programme at all levels.

(c) Efficient management of raw materials and waste products. Reduction of remediation cost and risk assessment. All these will help control product cost.

(d) **Institutional confidence** Implementation of ISO 14001 EMS enhances company's relationship with stakeholders, investors, banks and insurance companies.

(e) **Public relations** Implementation of ISO 14001 EMS will help meet environmental commitment to the local community.

14.6 URBAN AIR QUALITY MANAGEMENT

It consists of the following points

1. Air quality monitoring network.
2. Pollution emission inventories.
3. Numerical prediction models.
4. Air quality standards.
5. Public information systems.
6. Cost-effective pollution control policies and measures.

14.6.1 Objectives of air quality management

(a) To monitor and prevent pollution at sources.

(b) To encourage, develop and apply the best available technical solutions.

(c) To ensure that the polluter pays for the pollution and takes necessary steps to control pollution.

(d) To focus on protection on heavily polluted areas and riverbed.

(e) To devise air quality standards.

(f) To maintain vigilance and to anticipate future problems.

14.7 EPA-1986

This act is meant to control various environmental problems of India. The act consists of 26 sections described in four chapters regarding improvement of environment and its related matters.

The following powers are provided to the Central Government of India by EPA-1986

(a) Setting standards for various emission or discharge of environmental pollutant.

(b) Setting standards for the quality of environment in various aspects.

(c) Preparation of codes, guides and manuals relating to the abatement of environmental pollution etc.

(d) Planning and execution of a nation-wide programme for the prevention and control of environmental problems.

(e) Carrying out and sponsoring various investigations and research activities relating to the abatement of environmental pollution.

14.8 ORGANIZATION FOR ECONOMIC CO-OPERATION AND DEVELOPMENT (OECD)

In 1972, 24 countries established a forum to discuss the issues on international trades and environmental agreements and published a set of guiding principles concerning the international economic aspects of environmental policies. This forum is known as OECD. The OECD proposed the following four principles.

Principle 1 Polluter pays principle

The Government should have authority to implement necessary regulation to protect the environment. The polluter is the sole responsible for polluting the environment and should pay for that. However this principle has some drawbacks

1. Polluters are exempted after paying their fine. No other actions are taken against them.
2. It is difficult to develop a standard tax-chart for pollution because amount of fine must vary with type of pollution as well as degree of pollution.
3. Pollution taxes are largely influenced by politics.

Principle 2 Harmonization principle

All the Governments should follow the international environmental policies unless valid reasons for differences exist.

Principle 3 National Treatment and Non-discrimination principle

GATT principle of national treatment and non-discrimination should be applicable to all countries. According to this principle domestic and foreign products should not be discriminated by imposing taxes.

Principle 4 Compensating Import Levies and Export Rebates Principle

Countries should not try to neutralize the economic effect of differences in environmental policies by means of import duties and export rebates or equivalent measures.

Some international organizations for the protection and conservation of environment

(a) UNEP	United Nation Environment Programme, founded in 1972.
(b) EPA	Environmental Protection Agency, established in 1970.
(c) IUCN	International Union for Conservation of Nature (1948).
(d) CITES	Convention on International Trade in Endangered Species.
(e) UNESCO	United Nations Educational, Scientific and Cultural Organization.
(f) NEPA	National Environmental Policy Act.
(g) RCRA	Resource Conservation and Recovery Act.
(h) CAA	Clean Air Act.
(i) TSCA	Toxic Substances Control Act.

(j) GATT	General Agreement on Tariffs and Trade.
(k) IPR	Intellectual Property Right.

14.9 SOME IMPORTANT FEATURES OF EMS

14.9.1 Environmental policy and some key features of it

Definition Environmental policy is the key foundation with the help of which an organization interacts with its environment and takes its decisions for future policies regarding environmental concerns.

Key features of environmental policy are

(a) Activities, products, raw material resources, services and potential impacts of an organization on the environment.
(b) Communicated, implemented and maintained at all levels in the organization.
(c) Commitment to continual improvement of environmental performance.
(d) Setting of environmental objectives and statement, clearly indicating the steps to achieve the objectives.
(e) Environmental objectives and performance should be made public.

14.9.2 The guiding principles of environmental policy

Guiding principles are

(a) Adopt and aim to apply the principles of sustainable development.
(b) Adopt the highest available environmental standards in all site locations and to meet the applicable legislations.
(c) Adopt a total "Cradle-to-Grave" environmental assessment and accept responsibility for all products and services, the raw materials used and the wastes.
(d) Aim to minimize or eliminate the use of non-renewable resources and maximize the use of renewable resources.
(e) Accept strict liability for environmental damage.

14.9.3 Environmental policy statement

Environmental policy covers the overall environmental intentions and activities of an organization whereas environmental policy statement accurately reflects all aspects of an organization's environmental policy, indicating targets and objectives, relating to environmental matters.

Key features of environmental policy statement are

(a) Methods of minimizing impacts upon the natural environment.
(b) Waste management and energy efficiency.
(c) Review and reporting of environmental policies and practices.
(d) Environmental training.
(e) Public disclosure.
(f) Reduce consumption of non-renewable resources.

14.9.4 Environmental performance

Environmental performance of a firm is the extent and effectiveness of actions which the firm takes to mitigate the environmental consequences.

14.9.5 The principles of ISO 14001

The principles of ISO 14001 are

Principle 1 An organization should focus on its needs, ensure commitment to EMS and define its policy.

Principle 2 An organization should formulate a plan to fulfill its environmental policy.

Principle 3 An organization should develop the capabilities to achieve its environmental policy, objectives and targets.

Principle 4 An organization should measure, monitor and evaluate its environmental performance.

Principle 5 An organization should review and continually improve its EMS and environmental performance.

14.9.6 The Central Pollution Control Board (CPCB)

The full form of CPCB is Central Pollution Control Board. It is the national apex body for assessment, monitoring and control of water and air pollution. Some of the processes of monitoring environmental pollution in India are

1. Identification and mapping of industries, plants and vehicles using fossil fuels responsible for polluting environment by emitting GHG and other harmful gases.
2. Continuous checking of water quality of rivers and other water reservoirs.
3. Continuous monitoring of noise pollution in and around hospitals, schools and courts.
4. Studies of noise characteristics of various noise sources and its impact on human health.
5. Monitoring of ambient air quality at various selected stations.
6. Advise the Central Government on any matter concerning the improvement of the quality of air and the prevention, control or abatement of air pollution.
7. Organise through mass media a comprehensive programme regarding the prevention, control or abatement of air pollution.
8. Collect, compile and publish technical and statistical data relating to air pollution and the measures devised for its effective prevention, control or abatement and prepare manuals, codes, or guides relating to prevention, control or abatement of air pollution.
9. Lay down the standards for quality of air.

14.9.7 Different environmental problems in India

There are several environmental problems in India. These are listed below

(a) Water pollution and water scarcity.

(b) Air pollution.
(c) Soil pollution.
(d) Solid and hazardous wastes.
(e) Deforestation.
(f) Loss of biodiversity.
(g) Atmospheric changes.
(h) Population growth and poverty.

14.9.8 Different environmental legislations in India

Environmental legislations are

(a) Water (Prevention and Control of Pollution) Act (1974).
(b) Air (Prevention and Control of Pollution) Act (1981).
(c) Wildlife (Protection) Act (1972).
(d) Forest (Conservation) Act (1980).
(e) Environment (Protection) Act (1986).
(f) Hazardous Wastes (Management and Handling) Rules, 1989.
(g) Bio-Medical Waste (Management and Handling) Rules, 1998.
(h) Municipal Solid Wastes (Management and Handling) Rules, 2000.
(i) The Noise Pollution (Regulation and Control) Rules, 2000.
(j) Ozone Depleting Substances (Regulation) Rules, 2000.

14.10 ENVIRONMENTAL AUDIT

14.10.1 Introduction

Environmental auditing is a systematic, documented, periodic and objective process in assessing an organization's activities and services in relation to

(a) Assessing compliance with relevant statutory and internal requirements.
(b) Facilitating management control of environmental practices.
(c) Promoting good environmental management.
(d) Maintaining credibility with the public.
(e) Raising staff awareness and enforcing commitment to departmental environmental policy.
(f) Exploring improvement opportunities.
(g) Establishing the performance baseline for developing an Environmental Management System (EMS).

14.10.2 Types of environmental audit

There are three major types of audits

1. **Liability** Determines whether or not your company is in compliance with applicable laws and regulations.

2. **Management** Determines whether or not your company is meeting its stated goals.
3. **Activity Specific** Looks at the performance of your company in a specific issue area like carbon emissions or waste reduction.

14.10.3 Planning an environmental audit

1. Commitment

(a) Obtain commitment at the Directorate level.
(b) Communicate commitment to personnel at all levels.

2. Audit site and scope

(a) Audit site and boundary.
(b) Audit objective(s).
(c) Areas of audit.

14.10.4 Objectives of environmental audit

Audit objectives may contain the following

(i) Verification of legislative and regulatory compliance.
(ii) Assessment of internal policy and procedural conformance.
(iii) Establishment of current practice status.
(iv) Identification of improvement opportunities.

Areas of audit normally encompass

(i) Material management, savings and alternatives.
(ii) Energy management and savings.
(iii) Water management and economy of use.
(iv) Waste generation, management and disposal.
(v) Noise reduction, evaluation and control (internal and external).
(vi) Air emissions and indoor air quality.
(vii) Environmental emergency prevention and preparedness.
(viii) Transportation and travelling practices.
(ix) Staff awareness, participation and training in environmental issues.
(x) Environmental information publicity.
(xi) Public enquiry and complaints response.
(xii) Environmental management system set up, suitability and performance.

14.10.5 Environmental audit process

Pre-Audit

- Gain high level support.
- Determine the scope of the audit.

- Identify metrics and benchmarks.
- Establish a "no blame" policy.
- Select and authorize an audit team. If members of the team are to be comprised of consultants, choose and hire outside firm.
- Obtain relevant data.

Audit

- Review relevant company policies and procedures.
- Observe conditions during normal business hours.
- Perform appropriate tests.
- Interview employees.

Post-Audit

- Summarize and document findings.
- Make an action plan based on results for going forward.
- Inform relevant internal and external stakeholders of the results.
- Make plans for repeating the audit process in the future Resource Guide.

Final Audit Report addressing

- Environmental Legislation compliance status.
- Departmental environmental policies, procedures and guidelines conformity status.
- Status of current environmental performance.
- Recommendations for performance improvement.

14.10.6 Benefits of environmental audit

The major benefits of conducting an environmental audit include mitigating your company's legal and reputation risks, reducing operational inefficiencies, improving the environmental performance of your company to achieve ISO 14001 certifications.

(i) Risk mitigation

Environmental audits can act as an effective risk management tool for assessing compliance with environmental legislation and thereby assisting the company in avoiding the risk of prosecution and fines arising from violation of potential environmental laws. This is particularly true for those involved in hazardous polluting industries. Additionally, as a sustainable company, your stakeholders may have heightened expectations for your company's environmental performance. Conducting an environmental audit will help reassure stakeholders that company is living up to its environmental principles.

(ii) Reduction in operational inefficiencies

Environmental reporting can highlight areas of inefficiency in processes, for example, unnecessary waste of resources, product defects, excessive waste generation, etc. By identifying operational inefficiencies the company may be able to reduce costs, thereby improving company's environmental performance.

(iii) Achieving certification requirements

Conducting an environmental audit can be an important step to achieve ISO 14001 certifications or product specific certification from organizations like: Energy Star, Scientific Certification Services, etc.

14.11 LIFE CYCLE ASSESSMENT

14.11.1 Definition

LCA, also known as life cycle analysis, ecobalance, and *cradle-to-grave analysis*. The term 'life cycle' refers to the notion that a fair, *holistic* assessment requires

- the assessment of *raw material* production.
- manufacture of products.
- distribution of products.
- use of products.
- disposal of products.
- all intervening transportation steps necessary or caused by the product's existence.

A life cycle assessment is a technique to assess environmental impacts associated with all the stages of a product's life from-cradle-to-grave as mentioned earlier. LCA can help avoid a narrow outlook on environmental concerns by

(i) Compiling an inventory of relevant energy and material inputs and environmental releases.

(ii) Evaluating the potential impacts associated with identified inputs and releases.

(iii) Interpreting the results to help you make a more informed decision.

14.11.2 Goal

The goal of LCA is to compare the full range of environmental effects assignable to products and services in order to improve processes, support policy and provide a sound basis for informed decisions.

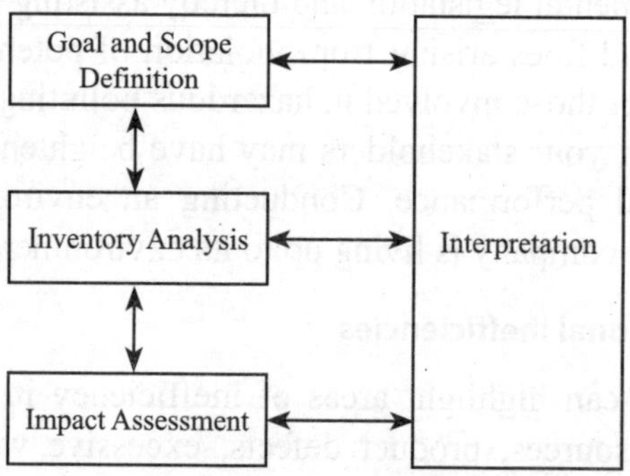

14.11.3 Goal and Scope

The goal and scope document includes technical details that guide subsequent work

(i) The functional unit, which defines what precisely is being studied and quantifies the service delivered by the product system, providing a reference to which the inputs and outputs can be related.

(ii) The system boundaries within which the functional unit is to be studied.

(iii) Assumptions and limitations to quantify the total unit. However, these assumptions and limitations vary from unit to unit.

(iv) Each and every unit has its own allocation methods in order to partition the environmental load of a process when several products or functions share the same process.

(v) The total system starting from raw material to product has different impacts on the environment and the overall impact depends on type of product and production rate. So impact category of different product is different.

14.11.4 Life Cycle Inventory

Life Cycle Inventory (LCI) analysis involves creating an inventory of flows from and to nature for a product system. Inventory flows include inputs of water, energy, and raw materials and releases to atmosphere, i.e. air, water and land.

14.11.5 Life Cycle Impact Assessment

Inventory analysis is followed by impact assessment. This phase of LCA is aimed at evaluating the significance of potential environmental impacts based on the LCI flow results. Classical Life Cycle Impact Assessment (LCIA) consists of the following mandatory elements

(i) Selection of impact categories, category indicators and characterization models.

(ii) The classification stage, where the inventory parameters are sorted and assigned to specific impact categories.

(iii) Impact measurement, where the categorized LCI flow is characterized, using one of many possible LCIA methodologies, into common equivalence units that are then summed to provide an overall impact category total.

14.11.6 Life Cycle Interpretation

Life Cycle Interpretation is a systematic technique to identify, quantify, check, and evaluate information from the results of the life cycle inventory and/or the life cycle impact assessment. According to ISO 14040:2006, the interpretation should include

(i) Identification of significant issues based on the results of the LCI and LCIA phases of an LCA.

(ii) Evaluation of the study considering completeness, sensitivity and consistency checks.

(iii) Conclusions, limitations and recommendations.

14.12 TECHNOLOGY DIFFUSION PROCESS

14.12.1 Definition

The diffusion process refers to the process of adoption of new products by the consumer marketplace. The consumer marketplace is segmented into four groups

(i) Innovators (2.5% of population).
(ii) Early adapters (13.5%), early majority (34%).
(iii) Late majority (34%).
(iv) Laggards (16%).

Thus innovators are more likely to adopt a new product earlier than the early majority, etc. This has implications for target marketing with new products. Very large investors belong to early majority, large investors belong to late majority category. Small investors generally do not take risk to adopt new product and hence they belong to Laggards category.

14.12.2 Technological progress

This is related to improvements in the techniques (including firm organization), related to production of goods, marketing and development. It is also related to the human progress and development. At the national level, technological progress can occur through

(i) Invention and innovation.
(ii) The adoption and adaptation of pre-existing but new-to-the-market technologies.
(iii) The spread of technologies across firms, individuals, and the public sector within a country.

The Fig. 14.3 illustrates the technology diffusion process.

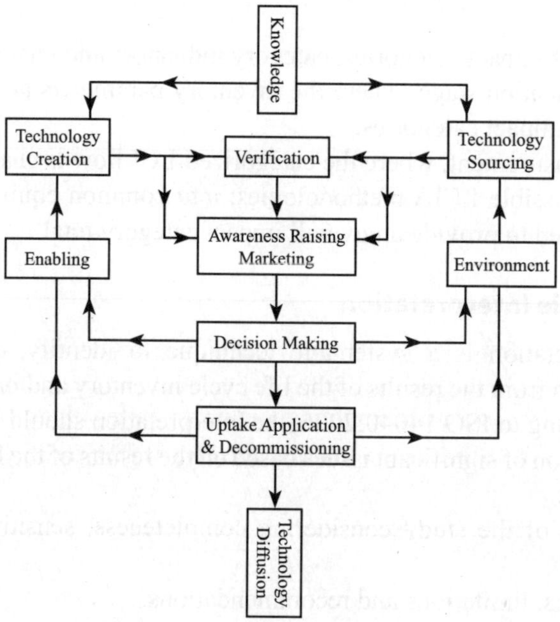

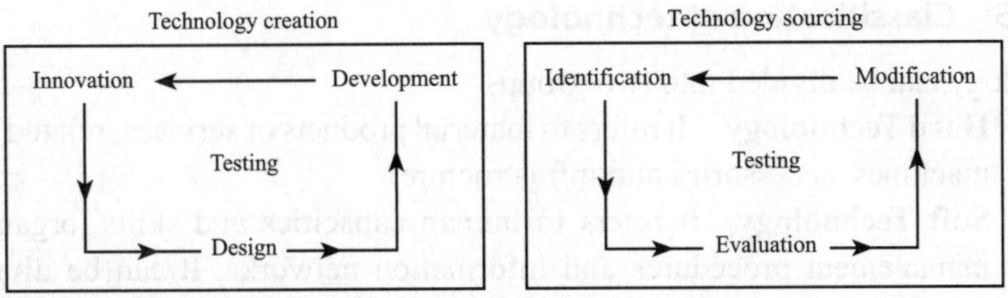

Fig. 14.3 Details of technology diffusion process.

14.12.3 Advantages of technology diffusion

(a) Technological progress can lower costs, improve quality, create new products, and help reach new markets.
  (i) Technological progress can spur development by lowering the costs of production and enabling the exploitation of increasing returns to scale.
  (ii) Technological progress in one sector can create new economic opportunities in other sectors.
  (iii) Technology can yield quality improvements.
  (iv) The benefits of a new technology can extend well beyond the immediate sector in which the technology exists.

(b) Even relatively simple technologies can have far-reaching development impacts.

(c) Technological change can also be disruptive. The disruptive nature of technological progress can generate important benefits to society by spurring competition. For example, the introduction of mobile phone technology in several developing countries has introduced an important element of competition not only in the telecommunications sector, but also in banking and other information-sensitive sectors.

(d) Scientific invention and innovation. Most technological improvements in developing countries are at least partially dependent on the diffusion of technology from more advanced countries.

(e) The intensity of innovation is closely related to per capita income.

14.12.4 The green revolution

The green revolution is an example of the dramatic effects that modest technological advances can have in boosting the incomes of the poor. The green revolution was a decade-long effort, guided primarily by public sector and nonprofit institutions, to create and disseminate agricultural technologies to developing countries. The principal technologies involved were pesticides, irrigation, and synthetic nitrogen fertilizer, which had long been available in industrial countries, along with the development of high-yielding varieties of maize, wheat, and rice.

14.12.5 Classification of technology

Technology can be divided into two groups

(a) **Hard Technology** It refers to material products or services, related to equipment, machines, accessories and infrastructure.

(b) **Soft Technology** It refers to human capacities and skills, organizational and management procedures and information networks. It can be divided into two subgroups

  (i) *Nature-based technology* natural process, mechanisms, materials, restoration capability.

  (ii) *Management tools* systems, procedures, guidelines, policy/regularity frameworks, capacity and capability enhancement.

The following are the most important points for Successful Transfer and Uptake of Environmentally Sound Technologies. As all these points start with C, it is popularly known as seven "C"s. The seven "C"s are

(i) Context,
(ii) Challenges,
(iii) Choice,
(iv) Certainty,
(v) Communication,
(vi) Capacity,
(vii) Commitment.

14.12.6 Conclusions

(a) Technology transfer must be recognised as a broad and complex process if it is to avoid creating and maintaining the dependency of the recipient, and if it is to contribute to sustained and equitable development.

(b) The end result for the recipient must be the ability to use, replicate, improve and, possibly, resell the technology.

(c) Transfer of technology is more than just moving of high-tech equipment and other so-called "hard" technologies from the developed countries to the developing or under developed countries. It also includes total systems and their component parts, including technological know-how, goods and services, equipment, organizational and managerial procedures.

(d) Thus technology transfer is the suite of processes encompassing all dimensions of the origins, flows and uptake of know-how, experience and equipment amongst, across and within countries, stakeholder organizations and institutions.

14.13 ENVIRONMENTAL IMPACT ASSESSMENT (EIA)

14.13.1 Introduction

14.13.1.1 Definition

It is the study of probable alterations in biophysical and socio-economic characteristics of the environment that may occur due to a proposed action, e.g. an industrial project.

14.13.1.2 Purposes

(a) Evaluation of the quality of the environment of the project site.
(b) Controlling the quality of the environment of the project site.
(c) Deriving alternative processes to maintain the quality of environment.

14.13.1.3 Parameters

The main purpose of EIA is to maintain the quality of environment and to sustain the ecosystem. It emphasises on the following points

(a) Alternate use of land-site.
(b) Alternate routes of development to manufacture the same product.
(c) Alternate technologies to derive other products.
(d) Strength of ecosystem to sustain the project.
(e) Quality of environment (before starting, during execution and after completion of the project).

14.13.1.4 Objectives

(a) Conservation of resources.
(b) Minimization of wastes.
(c) Recovery of by-products.
(d) Utilization of latest technology and efficient equipment.

14.13.2 Key-elements

(a) **Phase-I** Organising the job.
(b) **Phase-II** Assessment of performance.
(c) **Phase-III** Preparation of Environmental Impact Statement (EIS).
(d) **Phase-IV** Reviewing of Environmental Impact Statement.

14.13.2.1 Phase-I

An interdisciplinary (ID) team is formed with experts from several disciplines. The team collects data and information relevant to the project and analyses the same within a specified time frame.

14.13.2.2 Phase-II

(a) The ID team will visit the project site in person and assess the adverse effects of the project work on the environment.

(b) Existing environmental conditions of the project-site should be recorded.
(c) Alternate use of proposed project-site is to be considered.
(d) Alternate methods of achieving the same goal must also be studied.
(e) A check-list should be prepared to ensure complete coverage of all possible consequences.

14.13.2.3 Phase-III

All the data and information, collected from the project-site, is recorded in the form of a report, known as Environmental Impact Statement. It may contain the following

(a) Description of existing environment, pertaining to the proposed project.
(b) Detailed description of the proposed project (e.g. sources of raw materials, products, manpower, equipment, etc.)
(c) Description of the impact of the proposed project on the environment.
(d) Unavoidable adverse effects.
(e) Identification of remedial actions to be taken to minimize the above adverse effects.
(f) Alternate use of land-site.
(g) Alternate routes of development to achieve the same goal.
(h) Use of resources and recycle of resources.
(i) Incorporation of modifications in the proposed project.

14.13.2.4 Phase-IV

Reviewing the EIS The EIS report is reviewed in different steps as follows

(a) EIS must be presented to the public.
(b) A discussion session must be arranged among public representative, Government representative and project proposer.
(c) The abstract of the discussion should be annexed to the EIS report and the latter is submitted to the competent authority for review.
(d) After the final review of the environmental impacts and cost benefit analysis a decision is taken either to approve or to reject the project.

14.13.3 Benefits of EIA

The following are the benefits of implementation of EIA

(a) Avoiding violations of rules and regulations.
(b) Time and cost reduction for implementation of the project.
(c) Better performance of the project.
(d) Product cost reduction by implementing waste management technique.
(e) Improved, safe and better environment for living.
(f) Minimizing the use of resources.
(g) Maintenance of biodiversity.
(h) To increase community awareness and more knowledge about our environment.

15

Hardness of Water

15.1 WATER HARDNESS

Detergent forms lather or foam in presence of water. The main constituent of any detergent is the Na– or K– salt of fatty acids. The salt exists as colloids. These colloidal particles form micelles, each of which has two parts. One part remains dissolved in water, known as hydrophilic part and the other part repels water, known as hydrophobic part. This hydrophobic part forms lather and becomes negatively charged. These negatively charged lather attracts the dust and dirt from clothes and hence clothes are cleaned. Thus formation of lather is an important factor in washing clothes. Primary constituent of a micelle is $C_{17}H_{35}COONa$ or $C_{17}H_{35}COOK$. However, in presence of Ca^{2+}/Mg^{2+} ions the fatty acid salts get precipitated according to the following reactions

$$2C_{17}H_{35}COONa + CaCl_2 \longrightarrow (C_{17}H_{35}COO)_2Ca\downarrow + 2NaCl$$

$$2C_{17}H_{35}COONa + MgCl_2 \longrightarrow (C_{17}H_{35}COO)_2Mg\downarrow + 2NaCl$$

As a result of this cleaning efficiency of the detergent decreases. The phenomenon is known as hardening and the property of water is called hardness of water.

Definition of hardness of water It is the property of water to prevent forming lather or foam in presence of detergent. Besides Ca^{2+} and Mg^{2+} ions, other ions like Fe^{3+}, etc. also prevent forming lather but only Ca^{2+} and Mg^{2+} ions are present in significant amount in surface water and hence hardness of water mainly arises due to the presence of Ca^{2+} and Mg^{2+} ions.

15.2 TYPES OF HARDNESS

Hardness of water may be broadly classified into following categories

(i) Temporary hardness.
(ii) Permanent hardness.

As water inherently contains both type of hardness, total hardness of water is given by

Total hardness = Temporary hardness + Permanent hardness

15.2.1 Temporary hardness

15.2.1.1 Boiling method

When water sample is boiled, calcium and magnesium bicarbonates are precipitated as carbonates and hence hardness due to Ca– and Mg– bicarbonates is removed. This is called temporary hardness. So this temporary hardness mainly arises due to the presence of $Ca(HCO_3)_2$ and $Mg(HCO_3)_2$. These bicarbonates remain soluble in cold water but are removed as carbonates when boiled according to the following equation.

$$Ca(HCO_3)_2 \xrightarrow{\text{Heat}} CaCO_3\downarrow + CO_2 + H_2O$$

$$Mg(HCO_3)_2 \xrightarrow{\text{Heat}} MgCO_3\downarrow + CO_2 + H_2O$$

15.2.1.2 Clark's method

In this method temporary hardness can be removed through treatment of sample water by slaked lime, i.e. $Ca(OH)_2$. The following reactions occur during removal of temporary hardness

$$Ca(HCO_3)_2 + Ca(OH)_2 \xrightarrow{\text{Heat}} 2CaCO_3\downarrow + 2H_2O$$

$$Mg(HCO_3)_2 + Ca(OH)_2 \xrightarrow{\text{Heat}} CaCO_3\downarrow + MgCO_3\downarrow + 2H_2O$$

15.2.2 Permanent hardness

That fraction of hardness, which remains unaltered even after boiling, is known as permanent hardness. Presence of soluble salt of Ca^{2+} and Mg^{2+} ions other than bicarbonates, is the main source of permanent hardness.

15.2.3 Unit of hardness

The following units are used to express hardness of water

(i) *ppm*

(ii) *mg/l*

(iii) $\left(\frac{M}{100}\right)$

(iv) °Clarke = °Cl

(v) °French = °Fr

Out of these 5 different units, *ppm* is usually preferred in many cases. Other units can be expressed in terms of *ppm*. Hardness of water is generally expressed as parts per million or simply *ppm* of $CaCO_3$ equivalent. Molecular weight of $CaCO_3$ is 100. Taking this advantage hardness of water is usually expressed as $CaCO_3$ equivalent. So, water hardness 1 *ppm* means 10^6 ml sample water contains 1 gm $CaCO_3$ equivalent.

$$1\ ppm = \frac{1\,\text{gm}\ CaCO_3\ \text{equivalent}}{10^6\ \text{ml water}}$$

$$= \frac{10^3\ \text{mg}\ CaCO_3\ \text{equivalent}}{10^3\ \text{lit. of water}} = 1\ \text{mg/lit} = 10^{-3}\ \text{gm/lit}$$

Thus,
$$1000\ ppm = \frac{1\,\text{gm}\ CaCO_3\ \text{equivalent}}{1\ \text{lit. of water}}$$

$$= \frac{\left(\frac{1}{100}\right)\text{moles of}\ CaCO_3\ \text{equivalent}}{1\ \text{lit. of water}}$$

[Since, Molecular Weight of $CaCO_3$ is 100]

1 M = Molarity of $CaCO_3$ equivalent

$$= \frac{1\ \text{mole of}\ CaCO_3\ \text{equivalent}}{1\ \text{lit. of water}}$$

Thus,
$$1000\ ppm = 1 \times \left(\frac{M}{100}\right) CaCO_3\ \text{equivalent}$$

and
$$1\ ppm = 10^{-3}\left(\frac{M}{100}\right) CaCO_3\ \text{equivalent}$$

So, water hardness can also be expressed as $\left(\frac{M}{100}\right)$ $CaCO_3$ equivalent.

The different units of hardness can be expressed in terms of *ppm* as follows

$$1\ ppm = \frac{1\,\text{gm}\ CaCO_3\ \text{equivalent}}{10^6\ \text{ml water}}$$

$$1\ mg/l = 1\ ppm = \frac{1\,\text{gm}\ CaCO_3\ \text{equivalent}}{10^6\ \text{ml water}}$$

$$1 \times \left(\frac{M}{100}\right) = 10^3\ ppm = \frac{1\,\text{gm of}\ CaCO_3\ \text{equivalent}}{1\ \text{litre of water}}$$

$$1°Cl = 14.28\ ppm = \frac{64.8\,\text{mg}\,CaCO_3\ \text{equivalent}}{4.55\,\text{litres of distilled water}}$$

$$1°Fr = 10\ ppm = \frac{1\,\text{gm}\,CaCO_3\ \text{equivalent}}{10^5\ \text{ml of distilled water}}$$

15.2.4 Estimation of hardness of water using complexometric titration

15.2.4.1 Theory

Hardness of water arises mainly due to Ca^{2+} and Mg^{2+} salts. Estimation of hardness of a given sample water can be determined by using disodium salt of Ethylenediaminetetraacetic Acid, represented by Na_2EDTA. The latter is a strong complexing agent and easily forms complexes with Ca^{2+} and Mg^{2+}. The indicator, used for this purpose, is Eriochrome Black T (EBT), which itself is a complexing agent. When EBT is added to a given water sample, metal cations M^{2+} (Ca^{2+} and Mg^{2+}) form complexes with EBT according to the following equation

M^{2+} + EBT $\longrightarrow$ [M-EBT], which appears as wine red complex.

As this complex is stable at pH ≥ 10, ammonium chloride-ammonium hydroxide buffer solution is used to maintain the pH at around 10.

During titration with EDTA, EBT of the wine red complex ([M-EBT]) is replaced by EDTA to form ([M-EDTA]) since [M-EDTA] is more stable than [M-EBT]. The following reaction takes place

[M-EBT] + EDTA $\longrightarrow$ [M-EDTA] + EBT (sky blue).

Thus the change of wine red colour to distinct sky blue colour marks the end point of titration. Calculation of hardness from titration results gives total hardness.

To determine Ca–hardness separately, titration is to be carried out in presence of NaOH so that Mg^{2+} is bound as $Mg(OH)_2$, leaving behind free Ca^{2+} ions. Titration is to be carried out with Na_2EDTA as before by using murexide indicator instead of EBT indicator.

The formula of Ethylenediaminetetraacetic Acid is as below

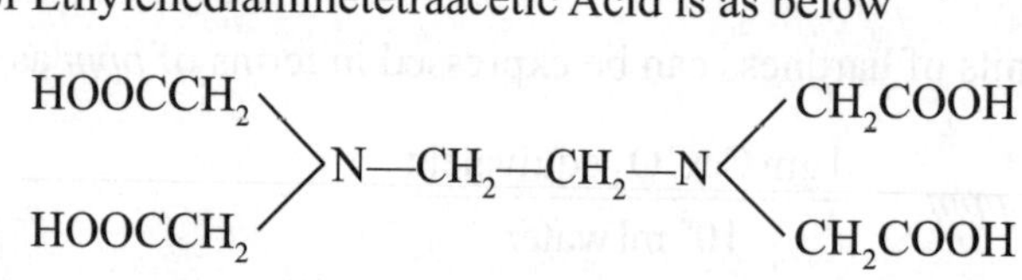

15.2.4.2 Procedure

(a) $Zn(CH_3COO)_2.2H_2O$ (MW = 219) solution of strength $0.5\left(\frac{M}{100}\right)$ is supplied.

(b) To prepare $0.5\left(\frac{M}{100}\right)$ Na_2EDTA solution around 200 mg Na_2EDTA crystals is dissolved in 100 ml distilled water in a 100 ml volumetric flask.

(c) Na_2EDTA solution is taken in a burette. 10 ml standard zinc acetate solution is taken in a 250 ml conical flask. 2 ml buffer solution is added to it followed by a pinch of indicator. The solution appears wine red, which is then titrated against Na_2EDTA solution until the colour changes to sky blue. Initial and final readings are noted and three such readings are recorded.

(d) In the next step 10 ml sample water is taken in a 250 ml conical flask. 2 ml buffer solution is added to it followed by a pinch of indicator. The solution appears wine red, which is then titrated against Na_2EDTA solution until the colour changes to sky blue. Initial and final readings are noted and three such readings are recorded.

(e) In the next step 10 ml sample water is taken in a 250 ml conical flask. 1 ml 2(N) NaOH solution is added to it to raise the *pH* to 12. A pinch of murexide indicator is added and the solution appears pink. The solution is then titrated against Na_2EDTA solution until the colour changes to purple. Initial and final readings are noted and three such readings are recorded.

15.2.4.3 Experimental Results

Table 15.1 Standardization of Na_2EDTA solution.

No. of observations	*Volume of standard zinc acetate solution taken (ml)*	*Volume of EDTA consumed (ml)*			
		Initial	*Final*	*Difference*	*Mean volume*
1	10				
2	10				V_2
3	10				

Calculation

$V_1S_1 = V_2S_2$, where,

V_1 = Volume of standard zinc acetate solution taken = 10 ml.

S_1 = Strength of standard zinc acetate solution = $0.5\left(\frac{M}{100}\right)$

V_2 = Mean burette reading of Table 15.1, in ml.

S_2 = Strength of Na_2EDTA solution.

Table 15.2 Estimation of total hardness of supplied hard water.

No. of observations	*Volume of sample hard water taken (ml)*	*Volume of EDTA consumed (ml)*			
		Initial	*Final*	*Difference*	*Mean volume*
1	10				
2	10				V_2
3	10				

Calculation

$V_1S_1 = V_2S_2$, where,

V_1 = Volume of sample hard water taken = 10 ml.

S_1 = Strength of sample hard water = $x\left(\frac{M}{100}\right)$

V_2 = Mean burette reading of Table 15.2, in ml.

S_2 = Strength of Na_2EDTA solution, already obtained from Table 15.1.

Water sample of strength $1 \times \left(\frac{M}{100}\right)$ is equivalent to water hardness of 1000 *ppm*.

So, water sample of strength $x\left(\frac{M}{100}\right)$ is equivalent to water hardness of 1000*x* *ppm*.

So, total hardness =1000*x* *ppm*.

Table 15.3 Estimation of Ca–hardness of the supplied hard water.

No. of observations	*Volume of sample hard water taken (ml)*	*Volume of EDTA consumed (ml)*			
		Initial	*Final*	*Difference*	*Mean volume*
1	10				
2	10				
3	10				

Calculation

$V_1S_1 = V_2S_2$

V_1 = Volume of sample hard water taken = 10 ml.

S_1 = Strength of sample hard water = $y\left(\frac{M}{100}\right)$

V_2 = Mean burette reading = ... ml.

S_2 = Strength of Na_2EDTA solution, already obtained from Table 15.1.

Water sample of strength $1 \times \left(\frac{M}{100}\right)$ is equivalent to water hardness of 1000 *ppm*.

So, water sample of strength $y\left(\frac{M}{100}\right)$ is equivalent to water hardness 1000*y* *ppm*.

So, total Ca–hardness = 1000*y* *ppm*.

So, total Mg–hardness = 1000 $(x - y)$ *ppm*.

15.2.4.4 Estimation of permanent hardness

200 ml of hard water sample is taken in a 500 ml beaker and heated to boil till the volume is reduced to one-half, i.e., 100 ml. The water is then cooled to room temperature. 50 ml

of that boiled water is taken in a 250 ml volumetric flask and filled up to the mark by distilled water. Then perform the procedure (d) mentioned above by taking 'boiled sample water' instead of 'sample water', calculate as before and find out the permanent hardness. Subtraction of permanent hardness from total hardness gives temporary hardness.

15.3 TREATMENT OF HARD WATER

Hard water should be treated in two steps:

15.3.1 External treatment

15.3.1.1 Lime-Soda process

The chemicals, used in this process, is a mixture of soda, i.e. Na_2CO_3 and slaked lime, i.e. $Ca(OH)_2$. The hard water and chemical mixture are heated to around 100°C with continuous stirring in a reaction chamber, where, the following reactions are believed to take place

$$CaCl_2 + Na_2CO_3 \longrightarrow CaCO_3\downarrow + 2NaCl$$

$$CaSO_4 + Na_2CO_3 \longrightarrow CaCO_3\downarrow + Na_2SO_4$$

$$MgCl_2 + Ca(OH)_2 \longrightarrow Mg(OH)_2\downarrow + CaCl_2$$

$$MgSO_4 + Na_2CO_3 + Ca(OH)_2 \longrightarrow Mg(OH)_2\downarrow + CaCO_3\downarrow + Na_2SO_4$$

As the process is carried out at high temperature, it is often called hot lime-soda process. Hardness of soft water usually lies within 30 *ppm*.

15.3.1.2 Zeolite process or Permutit process

In this process zeolite is used to remove the hardness of water. Its prime constituent is Na^+ ions and hence it is best represented as Na_2Ze. Its chemical formula is given below

$$\text{Zeolite} \equiv Ze = Na_2O{\cdot}Al_2O_3{\cdot}2SiO_2{\cdot}6H_2O$$

When hard water is passed over this Na_2Ze bed, Ca^{2+} and Mg^{2+} ions react with Na_2Ze, precipitating CaZe and MgZe. The following reactions are believed to take place

$$Na_2Ze + Ca(HCO_3)_2 \longrightarrow CaZe + 2NaHCO_3$$

$$Na_2Ze + Mg(HCO_3)_2 \longrightarrow MgZe + 2NaHCO_3$$

$$Na_2Ze + CaCl_2 \longrightarrow CaZe + 2NaCl$$

$$Na_2Ze + CaSO_4 \longrightarrow CaZe + Na_2SO_4$$

$$Na_2Ze + MgCl_2 \longrightarrow MgZe + 2NaCl$$

$$Na_2Ze + MgSO_4 \longrightarrow MgZe + Na_2SO_4$$

$$Na_2Ze + Fe(HCO_3)_2 \longrightarrow FeZe + 2NaHCO_3$$

$$Na_2Ze + Mn(HCO_3)_2 \longrightarrow MnZe + 2NaHCO_3$$

The process is illustrated in the Fig. 15.1.

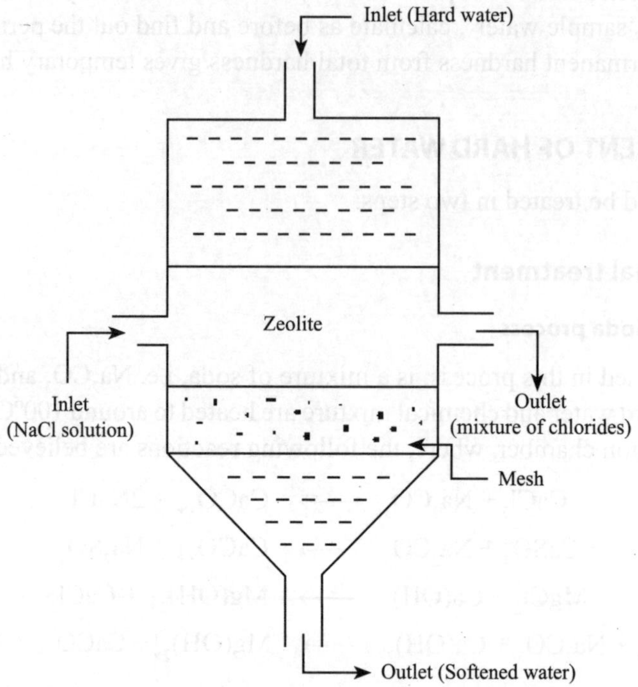

Fig. 15.1 Zeolite treatment of hard water.

The zeolite bed is regenerated by treating it with concentrated NaCl or brine solution. CaZe and MgZe react with NaCl to regenerate Na_2Ze.

Hardness of softened water lies < 10 *ppm*.

15.3.1.3 Demineralization or deionization process

In this process the raw hard water is passed through two successive filter beds. One is made of the resin, represented as $(RSO_3^- H^+)/(RCOO^-H^+)/RO^-H^+$, etc. which, when comes in contact with raw water, absorbs Ca^{2+}, Mg^{2+} or other cations with the release of H^+ ions.

$$nRSO_3^-H^+ + M^{n+} \longrightarrow (RSO_3)_nM + nH^+$$

This $RSO_3^-H^+$ is known as cation exchange resin and has the following chemical structure

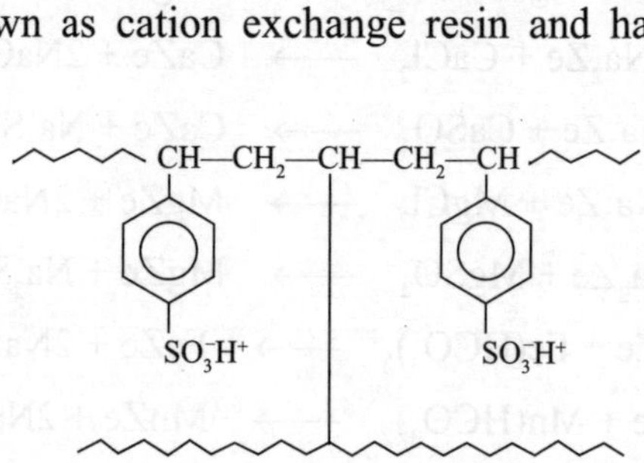

The effluent of cation exchange bed is sent to an anion exchange bed chamber, where all the anions are absorbed within the bed with the release of OH^- ions. The anion exchange resin is represented as $[RN(CH_3)_3]^+OH^-$. The chemical structure of the anion exchange resin is

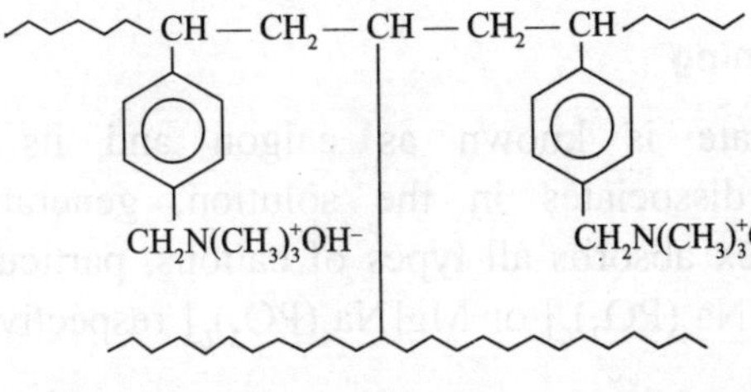

The following reaction is believed to take place during exchange of ions

$$n[RN(CH_3)_3]^+OH^- + A^{n-} \longrightarrow [RN(CH_3)_3]_nA + nOH^-$$

Finally H^+ and OH^- ions recombine to form pure water, which is known as demineralised water or deionized water.

$$H^+ + OH^- \longrightarrow H_2O$$

The hardness of deionized water is very low, as low as 1 *ppm*. The ion exchange bed are exhausted after certain time and then the beds can be regenerated by treating acid with cation exchange bed and alkali with anion exchange bed.

$$(RSO_3)_nM + nH^+ \longrightarrow nRSO_3^-H^+ + M^{n+}$$

$$[RN(CH_3)_3]_nA + nOH^- \longrightarrow n[RN(CH_3)_3]^+OH^- + A^{n-}$$

The process is widely used to soften any type of hard water. However, installation cost for setting up ion exchange plant is very high.

15.3.2 Internal treatment

Externally treated water is fed to the boiler but this water still possesses many undesired cations, particularly Ca^{2+} and Mg^{2+} ions, which form undesired scale and sludge. To eliminate these undesired cations the boiler feed water is treated again within the boiler. It is called internal treatment. It may be carried out in different ways as discussed below:

15.3.2.1 Phosphate conditioning

Sodium phosphate (Na_3PO_4) is used to remove undesired cations, i.e. Ca^{2+} and Mg^{2+}, from the boiler feed water. Both Ca^{2+} and Mg^{2+} ions form insoluble precipitates in presence of PO_4^{3-} ions. The following reactions are believed to take place

$$3CaCl_2 + 2Na_3PO_4 \longrightarrow Ca_3(PO_4)_2\downarrow + 6NaCl$$

$$3MgCl_2 + 2Na_3PO_4 \longrightarrow Mg_3(PO_4)_2\downarrow + 6NaCl$$

$$3CaSO_4 + 2Na_3PO_4 \longrightarrow Ca_3(PO_4)_2\downarrow + 3Na_2SO_4$$

$$3MgSO_4 + 2Na_3PO_4 \longrightarrow Mg_3(PO_4)_2\downarrow + 3Na_2SO_4$$

However, the above precipitation reactions occur at high alkaline medium (at pH ≥ 9.5). As Na_3PO_4 is highly alkaline, it is the most suitable reagent for treatment of water, having low alkalinity. Sometimes buffer solution (NH_4OH/NH_4Cl) is used to maintain pH at around 10. In that case Na_2HPO_4 and NaH_2PO_4 can be used instead of Na_3PO_4.

15.3.2.2 Calgon conditioning

Sodium hexametaphosphate is known as calgon and its chemical formula is $Na_2[Na_4(PO_3)_6]$. Calgon dissociates in the solution, generating the complex ion $[Na_4(PO_3)_6]^{2-}$. This complex absorbs all types of cations, particularly Ca^{2+} and Mg^{2+} to form soluble complex $Ca[Na_4(PO_3)_6]$ or $Mg[Na_4(PO_3)_6]$ respectively, thereby preventing scale and sludge formation.

15.4 ALKALINITY OF WATER

Concentration of acid required to neutralize a given sample water is known as alkalinity of water. It is expressed in terms of *ppm of $CaCO_3$ equivalent*. It mainly arises due to the presence of HCO_3^-, CO_3^{2-} and OH^-. These radicals are neutralized in presence of acid according to the following reactions

$$OH^- + H^+ \longrightarrow H_2O \qquad (1)$$

$$CO_3^{2-} + H^+ \longrightarrow HCO_3^- \qquad (2)$$

$$HCO_3^- + H^+ \longrightarrow H_2O + CO_2 \qquad (3)$$

As HCO_3^- is alkaline in nature, reactions (1) and (2) can be estimated by using phenolphthalein indicator. However, the reaction (3) can be estimated by using methyl orange only. Actually methyl orange indicator can be used to estimate all the above three reactions. Methyl orange is written as MO in short and phenolphthalein is written as P. CO_3^{2-} is neutralized in two steps [reactions (2) and (3)]. So reaction (2) indicates half neutralization of CO_3^{2-}. Thus we can write

$$P = [OH^-] + \frac{1}{2}[CO_3^{2-}]$$

and

$$MO = [OH^-] + [CO_3^{2-}] + [HCO_3^-] \qquad (4)$$

Where, P = Concentration of acid required in ml to neutralize a given sample water by using phenolphthalein as indicator.

MO = Concentration of acid required in ml to neutralize a given sample water by using methyl orange as indicator.

Several cases may arise depending on relations between P and MO.

Case-I: P = MO. It indicates from equation (4) that $[CO_3^{2-}] = [HCO_3^-] = 0$. Sample water contains only OH^- ions and $[OH^-] = P = MO$.

Case-II: P = 0. It indicates from equation (4) that $[CO_3^{2-}] = [OH^-] = 0$. Sample water contains only HCO_3^- ions and $[HCO_3^-] = MO$.

Case-III: $P = \frac{1}{2}MO$. It indicates from equation (4) that $[OH^-] = [HCO_3^-] = 0$. Sample water contains only CO_3^{2-} ions and $[CO_3^{2-}] = 2P = MO$.

Case-IV: $P > \frac{1}{2}MO$. It indicates from equation (4) that $[HCO_3^-] = 0$. Sample water contains OH^- and CO_3^{2-} ions. Thus we can write that

$$P = [OH^-] + \frac{1}{2}[CO_3^{2-}]$$

$$\frac{1}{2}MO = \frac{1}{2}OH^- + \frac{1}{2}[CO_3^{2-}]$$

$$P - \frac{1}{2}MO = \frac{1}{2}OH^-$$

Or, $OH^- = 2P - MO$

Again, $$P = [OH^-] + \frac{1}{2}[CO_3^{2-}]$$

$$MO = [OH^-] + [CO_3^{2-}]$$

$$MO - P = \frac{1}{2}[CO_3^{2-}]$$

$$[CO_3^{2-}] = 2(MO - P)$$

Case-V: $P > \frac{1}{2}MO$. It indicates from equation (4) that $[OH^-] = 0$. Sample water contains both HCO_3^- and CO_3^{2-} ions.

$$MO = [CO_3^{2-}] + [HCO_3^-]$$

$$P = \frac{1}{2}[CO_3^{2-}]$$

$$\frac{1}{2}MO - P = \frac{1}{2}[HCO_3^-]$$

$$[CO_3^{2-}] = 2P$$

Alkalinity attacks the boiler plate. Thus boiler feed water should be free from alkalinity. Alkalinity is usually expressed in terms of $CaCO_3$ equivalent.

Boiler feed water requirement

(a) Hardness should be less than 0.20 *ppm*.

(b) Alkalinity due to OH^- ion should be less than 0.40 *ppm*.

(c) Alkalinity due to CO_3^{2-} ion should be less than 1 *ppm*.

15.5 PROBLEMS ON BOILER

The following problems are frequently encountered with boiler

(a) Corrosion.
(b) Caustic embrittlement.
(c) Scale and sludge.
(d) Carry over.

15.5.1 Boiler corrosion

During steam formation at high pressure and temperature, boiler plate gets corroded or decayed due to several reasons. The following are believed to be the most responsible for boiler corrosion.

(i) Dissolved oxygen.
(ii) Carbon dioxide.
(iii) Mineral acids, specially, HCl.

(i) **Dissolved oxygen** Boiler feed water always contains some dissolved oxygen. Its concentration ranges from 7 *ppm* to 9 *ppm* or more. As solubility of oxygen decreases at high temperature, free oxygen is available near the boiler plate at high temperature. This free oxygen attacks the boiler plate to form rust on it.

$$\text{Boiler plate} + [O_2] \longrightarrow Fe(OH)_2 \longrightarrow Fe_2O_3.xH_2O \quad \text{(Rust)}$$

Thus free oxygen poses a threat to the boiler. To avoid this rust formation a reducing agent, preferably sodium sulphite (Na_2SO_3) is added. Whenever dissolved oxygen sets free, Na_2SO_3 reacts with O_2 to form Na_2SO_4, thereby protecting the boiler plate. The following reaction is believed to occur

$$2Na_2SO_3 + O_2 \longrightarrow 2Na_2SO_4$$

However, the excess Na_2SO_4 is also dangerous since it gives off corrosive sulphur di-oxide (SO_2) gas at high pressure. Thus very low concentration (as low as 1 *mg/l*) of Na_2SO_3 is used for this purpose.

(ii) **Carbon dioxide** CO_2 is often evolved during steam formation in boiler. At high pressure at boiler, bicarbonates are dissociated into carbonates with the release of CO_2. This CO_2 reacts with water to form corrosive H_2CO_3, which accelerates scale formation on the boiler plates. To avoid the formation of corrosive H_2CO_3, small amount of NH_4OH is added. This NH_4OH reacts with CO_2 to form ammonium carbonate [$(NH_4)_2CO_3$], which is soft and non-corrosive.

(iii) **Mineral acids** If boiler feed water contains significant amount of $MgCl_2$, the latter reacts with water to release corrosive HCl along with $Mg(OH)_2$, which forms scales on the boiler plates. HCl accelerates rust formation on the boiler plates. To avoid rust formation small amount of NH_4OH is added. This NH_4OH absorbs HCl before rust formation.

15.5.2 Caustic embrittlement

Boiler plates, after several hours of working, get corroded by caustic soda (NaOH) and become brittle. The phenomenon is called caustic embrittlement. This phenomenon is usually observed when water, softened by lime-soda process, is used at the boiler. This soft water often contains sufficient amount of sodium carbonate (Na_2CO_3), which at high pressure decomposes to give sodium hydroxide (NaOH) and carbon dioxide (CO_2). If the boiler plate contains multiple hair-cracks, NaOH flows through the hair-cracks and dissolves out iron from the boiler plate in the form of ferrates. Iron, wetted with NaOH solution, acts as anode and gets corroded continuously. CO_2 forms corrosive acid H_2CO_3 which is also dangerous to boiler plate.

$$Na_2CO_3 + H_2O \longrightarrow 2NaOH + CO_2$$

As a result of this, stress corrosion develops on the boiler plate, leading to boiler failure.

Use of zeolite-softened soft water also leads to caustic embrittlement since $NaHCO_3$ is produced in huge amount during zeolite softening process. This $NaHCO_3$ at high temperature decomposes to sodium hydroxide (NaOH) and carbon dioxide (CO_2).

$$NaHCO_3 \longrightarrow NaOH + CO_2$$

Prevention It is evident that caustic embrittlement occurs due to carbonates and bicarbonates. Thus sample water should be boiled for long time to remove temporary hardness. After that zeolite is used to remove permanent hardness. Finally small amount of Na_2SO_4 is added. This Na_2SO_4 fills up pores and cracks of boiler so that NaOH does not get in contact with boiler plate.

If Na_3PO_4 is used to remove permanent hardness, the following reactions may occur:

$$2CaSO_4 + 2Na_3PO_4 \longrightarrow 2Na_2SO_4 + Ca_3(PO_4)_2\downarrow$$

$$3MgSO_4 + 2Na_3PO_4 \longrightarrow 2Na_2SO_4 + Mg_3(PO_4)_2\downarrow$$

In this case there is no need to add Na_2SO_4 in boiler feed water, since Na_2SO_4 is already produced in the reactions.

15.5.3 Scale and Sludge

The function of a boiler is to supply steam continuously. Water is continuously supplied to the boiler, at which water is converted to steam at high temperature and pressure. During boiling, impurities, present in water, gets deposited on the boiler plate continuously.

Some precipitates are very fine and remain within boiler in the form of colloids. This is called sludge. The main constituents of sludge are $CaCl_2$, $MgCl_2$, $MgCO_3$, $MgSO_4$, etc.

Some precipitates get precipitated in the form of coarse particles, which remain adhered with inner wall of the boiler plate as hard deposit. This is called scale. The main constituents of scale are $CaCO_3$, $CaSO_4$, $Mg(OH)_2$, $CaSiO_3$, and $MgSiO_3$.

Disadvantages of scale and sludge formation

(a) Usually scales and sludge are poor conductor of heat and hence excess fuel is often required to achieve the desired temperature. Thus fuel wastage is the immediate effect of scale and sludge formation.
(b) Boiler efficiency decreases.
(c) Corrosion of boiler plate.
(d) Overheating of boiler plate. This is a serious problem since due to overheating high heat is developed on the boiler plate, as a result of which, metal becomes weaker, leading to the formation of cracks on the boiler plate.
(e) As thermal conductivity of boiler is adversely affected due to scale and sludge formation, some portion of boiler remains over-heated whereas other portion remains under-heated. As a result of this, boiler tube gets distorted, thereby decreasing the boiler efficiency and threatening the danger of explosion.

Prevention To avoid the scale and sludge formation, soft water, having following specifications, is recommended for the use in the boiler.

(a) Hardness should be less than 0.20 *ppm*.
(b) Alkalinity due to OH^- ion should be less than 0.40 *ppm*.
(c) Alkalinity due to CO_3^{2-} ion should be less than 1 *ppm*.
(d) Very pure form of soft water.

15.5.4 Carry over

During steam generation in a boiler, two phenomena, called priming and foaming, happen to take place very often.

Water droplets are often associated with the steam, generated in the boiler. This steam is called wet steam and the phenomenon of forming wet steam is called priming. The water droplets in the wet steam may possesses impurities, which decrease the efficiency of the boiler and also lead to corrosion. The impurities are carried by the steam to the turbine and get deposited there. This undesired deposits damage the turbine.

The boiler feed water often contains oil and alkalis, which react with each other during steam generation at high temperature and pressure, generating foam. The phenomenon is called foaming. Foaming is highly undesirable since it generates numerous air pockets and changes the boiler pressure. Addition of small amount of castor oil and antifoaming agent, e.g. $NaAlO_2, Al_2(SO_4)_3$, etc. prevents the foaming.

15.6 NUMERICAL PROBLEMS ON WATER SOFTENING BY LIME-SODA PROCESS

$Ca(HCO_3)_2 + L \longrightarrow 2CaCO_3\downarrow + H_2O$ [Only Lime is required] L

$Mg(HCO_3)_2 + 2L \longrightarrow 2CaCO_3\downarrow + Mg(OH)_2\downarrow + 2H_2O$

[Only Lime is required] 2L

$(MgCl_2)/(MgSO_4) + L \longrightarrow Mg(OH)_2\downarrow + Ca^{2+}$

$Ca^{2+} + S \longrightarrow CaCO_3\downarrow + 2Na^+$ [Both Lime and Soda are required] (L + S)

$(CaCl_2)/(CaSO_4) + S \longrightarrow CaCO_3\downarrow + 2Na^+$ [Only Soda is required] S

$CO_2 + L \longrightarrow CaCO_3\downarrow + H_2O$ [Only Lime is required] L

$2H^+ + L \longrightarrow Ca^{2+} + 2H_2O$

$Ca^{2+} + S \longrightarrow CaCO_3\downarrow + 2Na^+$ [Both Lime and Soda are required] (L + S)

$2HCO_3^- + L \longrightarrow CaCO_3\downarrow + S + H_2O$ [Soda is produced] (L – S)

$FeSO_4 + L \longrightarrow Fe(OH)_2 + CaSO_4 \longrightarrow Fe(OH)_3\downarrow$

$CaSO_4 + S \longrightarrow CaCO_3\downarrow + 2Na^+$ [Both Lime and Soda are required] (L + S)

$Al_2(SO_4)_3 + 3L \longrightarrow 2Al(OH)_3\downarrow + 3CaSO_4$

$3CaSO_4 + 3S \longrightarrow 3CaCO_3\downarrow + 3Na_2SO_4$

[Both Lime and Soda are required] (3L + 3S)

Chemical formula	*Molecular weight*
Lime $[Ca(OH)_2]$	74
Soda $[Na_2CO_3]$	106
$MgSO_4$	120
$MgCO_3$	84
$Mg(HCO_3)_2$	146
$MgCl_2$	95
$Ca(HCO_3)_2$	162
$CaCO_3$	100
$CaSO_4$	136
$CaCl_2$	111

Problem 1 Analysis of a sample water gave the following result:

$H_2SO_4 = 200$ *ppm*, $MgSO_4 = 30$ *ppm*, $CaSO_4 = 300$ *ppm*, NaCl = 35 *ppm*

Calculate the amount of lime and soda required to soften 150×10^6 *l* of water.

Solution Lime requirement

$$= \frac{74}{100}\{[H_2SO_4] + [MgSO_4]\}CaCO_3 \text{ eq. } (mg/l) \times 150 \times 10^6\, l$$

$$= \frac{74}{100}\left\{200 \times \frac{100}{98} + 30 \times \frac{100}{120}\right\} \times 150 \text{ kg.}$$

= 25428 kg (approx.).

Soda requirement

$$= \frac{106}{100}\{[H_2SO_4] + [MgSO_4] + [CaSO_4]\}CaCO_3 \text{ eq. } (mg/l) \times 150 \times 10^6\ l$$

$$= \frac{106}{100}\left\{200 \times \frac{100}{98} + 30 \times \frac{100}{120} + 300 \times \frac{100}{136}\right\} \times 150 \text{kg.}$$

= 71497.5 kg.

Problem 2 Calculate the amount of lime and soda required for softening $2 \times 10^6\ l$ of water containing the following impurities:

H^+ : 2 *ppm*, HCO_3^- : 400 *ppm*, Mg^{2+} : 45 *ppm*, Ca^{2+} : 95 *ppm*, $FeSO_4.7H_2O$: *ppm*.

Problem 3 Calculate the amount of lime and soda required for softening $3 \times 10^6\ l$ of water containing the following impurities:

HCl : 4 *ppm*, $FeCl_3$: 10 *ppm*, $MgSO_4$: 6 *ppm*, $Mg(HCO_3)_2$: 3 *ppm*, $Ca(HCO_3)_2$: 3 *ppm*, $CaSO_4$: 18 *ppm*.

Problem 4 Calculate the amount of lime and soda required for softening $4 \times 10^6\ l$ of water containing the following impurities:

Ca^{2+} : 180 *ppm*, HCO_3^- : 395 *ppm*, Mg^{2+} : 100 *ppm*, CO_3^{2-} : 200 *ppm*, CO_2 : 30 *ppm*.

Problem 5 Calculate the quantity of lime and soda required for treatment of $2 \times 10^6\ l$ of water, having the following impurities:

$Ca(HCO_3)_2$: 40 *ppm*, $MgSO_4$: 28 *ppm*, $Mg(HCO_3)_2$: 42 *ppm*

$CaSO_4$: 32 *ppm*, $CaCl_2$: 24 *ppm*, NaCl : 12 *ppm*.

Problem 6 Calculate the amount of lime and soda required for softening $2 \times 10^6\ l$ of hard water, which was analyzed as follows:

Substance	*Concentration (mg/l)*
$Ca(HCO_3)_2$	243
$Mg(HCO_3)_2$	73

$CaSO_4$	102
$MgCl_2$	95
NaCl	500
$FeSO_4.7H_2O$	139

Purity of lime is 94% and that of soda is 86%. 10% chemicals are used in excess in order to complete the reaction quickly.

15.7 NUMERICAL PROBLEMS ON ALKALINITY

Problem 1 100 ml of a water sample consumes 7 ml $\left(\frac{N}{50}\right)$ H_2SO_4 up to the end point, using phenolphthalein as indicator but 100 ml of same water sample consumes 14 ml $\left(\frac{N}{50}\right)$ H_2SO_4 up to the end point, using methyl orange as indicator. Determine the type of alkalinity and calculate the alkalinity of the water sample.

Solution **Case-I** Titration with phenolphthalein indicator.

We know: $V_1S_1 = V_2S_2$, where,

V_1 = Volume of water sample = 100 ml.

S_1 = Strength of water sample, which is to be determined.

V_2 = Volume of H_2SO_4 consumed = 7 ml.

S_2 = Strength of H_2SO_4 = (N/50).

Hence, $S_1 = 0.07\left(\frac{N}{50}\right) = 0.07\left(\frac{M}{50}\right)$ equivalent $CaCO_3$.

since Mol. wt. of $CaCO_3$ = 2 × Eq. wt. of $CaCO_3$

Or, $1\,N\,CaCO_3 = \left(\frac{M}{2}\right) CaCO_3$

Again, $1 \times \left(\frac{M}{100}\right) = 1000\ ppm$

So, $P = 0.07\left(\frac{M}{100}\right) = 70\ ppm$

Case-II Titration with methyl orange indicator.

We know: $V_1S_1 = V_2S_2$, where,

V_1 = Volume of water sample = 100 ml.

S_1 = Strength of water sample, which is to be determined.

V_2 = Volume of H_2SO_4 consumed = 14 ml.

S_2 = Strength of H_2SO_4 = $\left(\frac{N}{50}\right)$.

Hence, $$S_1 = 0.14\left(\frac{N}{50}\right) = 0.14\left(\frac{M}{100}\right) \text{ equivalent } CaCO_3$$

since Mol. wt. of $CaCO_3$ = 2 × Eq. wt. of $CaCO_3$

Or, $$1\,N\,CaCO_3 = \left(\frac{M}{2}\right) CaCO_3$$

Again, $$1 \times \left(\frac{M}{100}\right) = 1000\,ppm$$

So, $$MO = 0.14\left(\frac{M}{100}\right) = 140\,ppm$$

Thus, $$P = \frac{1}{2}\text{MO}$$

So, only CO_3^{2-} ions are present in the given water sample.

Problem 2 100 ml of a water sample consumes 6 ml $\left(\frac{N}{50}\right)$ HCl up to the end point, using phenolphthalein as indicator but 100 ml of same water sample consumes 16 ml $\left(\frac{N}{50}\right)$ HCl up to the end point, using methyl orange as indicator. Determine the type of alkalinity and calculate the alkalinity of the water sample.

Problem 3 A sample is alkaline both to phenolphthalein and methyl orange. 50 ml of this water sample requires 15 ml $\left(\frac{N}{50}\right)$ H_2SO_4 up to the end point, using phenolphthalein as indicator and another 10 ml for complete neutralization. Calculate the type of alkalinity.

Problem 4 A water sample is not alkaline to phenolphthalein. 100 ml of this sample requires 18 ml $\left(\frac{N}{50}\right)$ HCl up to the end point, using methyl orange as indicator. Determine the type of alkalinity and amount of alkalinity present in the sample.

Problem 5 100 ml of a water sample consumes 6 ml $\left(\frac{N}{50}\right)$ H_2SO_4 up to the end point, using phenolphthalein as indicator. 100 ml of same water sample consumes 6 ml $\left(\frac{N}{50}\right)$ H_2SO_4 up to the end point, using methyl orange as indicator. Determine the type of alkalinity and calculate the alkalinity of the water sample.

Problem 6 200 ml of a water sample consumes 20 ml $\left(\frac{N}{50}\right)$ HCl using methyl orange as indicator. Another 200 ml of same sample requires 8 ml $\left(\frac{N}{50}\right)$ HCl using phenolphthalein as indicator. Calculate the alkalinity of the water sample in terms of $CaCO_3$ equivalent in *mg/l*.

Appendix-I

Table 1 Physical Constants

Avogadro's Number	N	6.023×10^{23}/g-mole or 6.023×10^{26}/kg-mole
Molar Gas Constant	R	8.314 J/(mole-°K) = 1.98 cal/(mole-°K) = 0.082 lit-atm/(mole-°K)
Boltzmenn's Constant	$k_B = R/N$	1.38×10^{-23} J/°K
Faraday's Constant	F	96,500 C/gm-equivalent of element
Electronic Charge	$e = F/N$	1.602×10^{-19} C $= 4.806 \times 10^{-10}$ e.s.u. of charge.
Atomic mass unit	amu = $1/N$	1.66×10^{-24} gm = 1.66×10^{-27} kg
Velocity of light	C	3.0×10^{8} m/s.
Electron rest mass	m_o	9.11×10^{-31} kg.
Bohr magneton (B.M.)	μ_B	9.273×10^{-24} A-m^2
Permittivity of free space	ε_o	8.854×10^{-12} F/m
Acceleration due to gravity	g	9.81 m/s^2
Dipole moment (1 Debye)	D	10^{-18} e.s.u.-cm
Gravitational Constant	G	6.6732×10^{-11} N-m^2/kg^2
Stefan–Boltzmann Constant	σ	5.67×10^{-8} W/(m^2-°K^4)
Rydberg Constant	R_B	109737 cm^{-1}

SI unit prefixes: 1 kilo = 10^3 unit, 1 mega = 10^6 unit, 1 giga = 10^9 unit, 1 tera = 10^{12} unit, 1 milli = 10^{-3} unit, 1 micro = 10^{-6} unit, 1 nano = 10^{-9} unit, 1 pico = 10^{-12} unit, 1 femto = 10^{-15} unit, 1 atto = 10^{-18} unit, 1⊕ = 10^{-10} m.

Table 2 Conversion Factors

Length:	1 m = 3.28, ft = 1.094 yd. 1 in = 2.54 cm, 1 ft = 30.48 cm. 1 mile = 1.6 km.
Area:	1 m^2 = 10.764 ft^2 1 in^2 = 645.16 mm^2 1 hectare = 100 Ares = 10,000 m^2 = 2.471 acres.
Force:	1 N = 0.2248 lbf, 1 N = 10^5 dyne 1 kgf = 9.81 N
Energy:	1 J = 0.738 ft-lbf = 10^7 erg 1 eV = 1.602 $\times 10^{-19}$ J 1 cal = 4.19 J 1 J = 0.2392 Cal = 6.242 $\times$ 10^{18} eV 1 B.T.U. = 252 Cal.
Volume:	1 ft^3 = 28316.85 cm^3 or c.c. = 7.48 gallon (U.S.) 1 gallon = 3786 c.c. 1 m^3 = 35.288 ft^3 1 barrel = 42 US-gallon
Pressure:	1 torr = 1 mm Hg at 0°C 1 atm = 101.325 kPa = 14.7 psi = 1.01325 bar = 760 mm Hg at 0°C 1 bar = 100 kPa = 10^5 Pa 1 kgf/cm^2 = 9.810 $\times$ 10^4 Pa = 0.981 bar = 0.968 atm. 1 Mpa = 145.1 psi = 1 N/mm^2 = 9.87 atm = 10.06 kgf/cm^2. 1 tsi = 15.2 Mpa
Temperature:	t°F = t°C + 32 T (R) = t°F + 459.67 T (K) = t°C + 273 T (R) = 1.8 T (K)
Power:	1 kW = 239.2 cal/s = 1.341 hp. 1 hp = 745 W = 0.745 kW 1 lumen (lm) = 1.496 $\times$ 10^{-10} W 1 lm/ft^2 = 1 foot-candle (fc) = 1.609 $\times$ 10^{-12} W/m^2 1 lm/ft^2 = Luminous flux = 1 foot–candle (fc)
Mass and density:	1 kg = 1000 gm = 2.2046 lb. 1 g/cm^2 = 1000 kg/m^3 = 62.43 lb/ft^3
Coefficient of Viscosity:	1 poise (p) = 100 centi-poise (cp) = 0.1 Pa-sec. 1 centi-poise(cp) = 10^{-3} Pa-sec = 6.7197 $\times$ 10^{-4} lb/ft-s = 2.4191 lb/ft-h.

Table 3(a) Standard Oxidation Electrode Potentials (E^0_{298}) of selected redox systems in aqueous solution.

Reduced Form	*Oxidized Form*	*No. of electron(s) released*	E^0_{298} *(V)*
Li	Li^+	1*e*	+ 3.05
K	K^+	1*e*	+ 2.936
Ca	Ca^{2+}	2*e*	+ 2.87
Na	Na^+	1*e*	+ 2.71
Mg	Mg^{2+}	2*e*	+ 2.36
Al	Al^{3+}	3*e*	+ 1.67
Mn	Mn^{2+}	2*e*	+ 1.18
½ $H_2(g) + OH^-$	H_2O	1*e*	+ 0.828
Zn	Zn^{2+}	2*e*	+ 0.76
Fe	Fe^{2+}	2*e*	+ 0.44
Cd	Cd^{2+}	2*e*	+ 0.40
$Pb + 2I^-$	PbI_2	2e	+ 0.365
$Pb + SO_4^{2-}$	$PbSO_4$	2*e*	+ 0.356
Sn	Sn^{2+}	2*e*	+ 0.14
Pb	Pb^{2+}	2*e*	+ 0.126
Fe	Fe^{3+}	3*e*	+ 0.04
½ D_2	D^+	1*e*	+ 0.01
½ H_2	H^+	1*e*	0.00
Cu	Cu^{2+}	2*e*	– 0.34
Cu	Cu^+	1*e*	– 0.518

Table 3(b) Standard Oxidation Electrode Potentials (E^0_{298}) of selected redox systems in aqueous solution.

Reduced Form	*Oxidized Form*	*No. of electron(s) change*	E^0_{298} *(V)*
I^-	½ I_2	1*e*	– 0.54
2 Hg(l) + SO_4^{2-}	Hg_2SO_4	2*e*	– 0.615
Fe^{2+}	Fe^{3+}	1*e*	– 0.77
Ag	Ag^+	1*e*	– 0.80
Hg	Hg^{2+}	2*e*	– 0.85
NO + $2H_2O$	NO_3^- + $4H^+$	3*e*	– 0.96
Br^-	½ Br_2	1*e*	–1.07
Mn^{2+} + $2H_2O$	MnO_2 + $4H^+$	2*e*	– 1.23
H_2O	½ O_2(g) + $2H^+$	2*e*	– 1.23
$2Cr^{3+}$ + $7H_2O$	$Cr_2O_7^{2-}$ + $14H^+$	6*e*	– 1.33
Cl^-	½ Cl_2	1*e*	– 1.36
Cl^- + $3H_2O$	ClO_3^- + $6H^+$	6*e*	– 1.45
Mn^{2+} + $4H_2O$	MnO_4^- + $8H^+$	5e	– 1.51
Ce^{3+}	Ce^{4+}	1*e*	– 1.61
F^-	½ F_2	1*e*	– 2.87
Ag + Br^-	AgBr	1*e*	– 0.073
SO_3^{2-} + H_2O	SO_4^{2-} + $2H^+$	2*e*	– 0.22
Ag + Cl^-	AgCl	1*e*	– 0.2222
2Hg(l) + $2Cl^-$	Hg_2Cl_2	2*e*	– 0.2680

Table 4 Typical oxyacids with their acid strength.

Acid	*Ka*	*pKa*	*Acid strength*
HOBr	2.06×10^{-9}	8.7	Very Weak
HOCl	3.2×10^{-8}	7.5	Very Weak
HOI	2.3×10^{-11}	10.64	Very Weak
H_3AsO_3	6.0×10^{-10}	9.2	Very Weak
H_3BO_3	5.8×10^{-10}	9.24	Very Weak
H_4SiO_4	2.0×10^{-10}	9.7	Very Weak
H_6TeO_6	2.0×10^{-8}	7.7	Very Weak
$HClO_2$	1.1×10^{-2}	1.97	Weak
HNO_2	4.0×10^{-4}	3.40	Weak
H_2CO_3	1.32×10^{-4}	3.88	Weak
H_2SO_3	1.3×10^{-2}	1.87	Weak
H_3AsO_4	6.46×10^{-3}	2.19	Weak
H_5IO_6	3.1×10^{-2}	1.57	Weak
H_3PO_4	7.25×10^{-3}	2.12	Weak
$HBrO_3$	2.0×10^{-1}	0.7	Strong
$HClO_3$	10	– 1	Strong
HIO_3	1.7×10^{-1}	0.77	Strong
HNO_3	43.6	– 1.64	Strong
H_2MnO_4	10^{-1}	1.0	Strong
H_2SO_4	1.0×10^{3}	– 3.0	Very Strong
$HClO_4$	10^{10}	– 10.0	Very Strong
$HMnO_4$	2.0×10^{2}	– 2.3	Very Strong

Table 5 (a) Coefficient of viscosities (η) of liquids in centi-poise (*cp*) at 20°C.

Liquid	*η*	*Liquid*	*η*
Water	1.005	Ethyl ether	0.245
Ethanol	1.194	Chloroform	0.563
Methanol	0.593	Chlorobenzene	0.637
Acetone	0.331	Benzene	0.647
Carbon tetrachloride	0.958	Acetic acid	1.222
Nitrobenzene	1.98	Glycerin	10.69

Table 5 (b) Coefficient of viscosities (η) of several liquid metals and alloys.

Material	*Temperature (°C)*	*Viscosity (η) (cp)*
Aluminium	700	3.0
Copper	1200	3.2
Iron	1600	6.2
White cast iron	1300	2.4
Steel	> M.P.	~ 3.0
Magnesium	680	1.2
Zinc	500	3.7

Table 6 Solubility products of some barely soluble compounds in water at 25°C.

Compound	*Solubility product (Ksp)*	*Compound*	*Solubility product (Ksp)*
AgBr	5.3×10^{-13}	AgCN	1.4×10^{-16}
AgCl	1.78×10^{-10}	Ag_2CrO_4	2.5×10^{-12}
AgI	8.30×10^{-17}	PbS	2.5×10^{-27}
$CaSO_4$	9.1×10^{-6}	CdS	4.0×10^{-29}
$SrSO_4$	2.8×10^{-7}	CuS	3.0×10^{-42}
$BaSO_4$	1.1×10^{-10}	HgS	3.0×10^{-54}
$PbCl_2$	2.4×10^{-4}	CoS	2.0×10^{-27}
$PbBr_2$	7.9×10^{-5}	ZnS	1.0×10^{-23}
PbI_2	8.7×10^{-9}	NiS	1.4×10^{-24}
$CaCO_3$	4.8×10^{-9}	MnS	2.0×10^{-15}
$BaCO_3$	5.1×10^{-9}	$Mg(OH)_2$	6.0×10^{-13}
$Ca_3(PO_4)_2$	2.0×10^{-29}	CaF_2	4.0×10^{-11}
$AlPO_4$	5.75×10^{-19}	CuI	5.0×10^{-12}
CuBr	1.60×10^{-11}	CuCl	1.0×10^{-6}

Table 7 Surface tension values of liquid metals and other liquids.

Material	*Temperature (°C)*	*Surface tension (γ) (Dynes/cm)*
Aluminium	750	520
Copper	1200	1160
Iron	1600	1360
Lead	350	453
Magnesium	681	563
Mercury	20	465
Zinc	600	770
Fused salts	400 to 930	35 to 180
Water	20	74

Table 8 Values of C_p, C_v and γ for ideal gases at 25°C.

Gas	*Chemical formula*	*C_p(kJ/kg-°K)*	*C_v(kJ/kg-°K)*	*γ*
Air	—	1.0035	0.7165	1.4
Oxygen	O_2	0.922	0.622	1.393
Nitrogen	N_2	1.042	0.745	1.4
Argon	Ar	0.520	0.312	1.667
Hydrogen	H_2	14.209	10.085	1.409
Helium	He	5.193	3.116	1.667
Carbon monoxide	CO	1.041	0.744	1.4
Carbon dioxide	CO_2	0.842	0.653	1.289
Methane	CH_4	2.254	1.736	1.299
Ethane	C_2H_6	1.7662	1.4897	1.86
Propane	C_3H_8	1.6794	1.4909	1.126
Isobutene	C_4H_{10}	1.716	1.573	1.091
Acetylene	C_2H_2	1.712	1.394	1.23
Water vapour (Steam)	H_2O	1.8723	1.4108	1.327

Table 9 Average bond energy values in kJ/mole.

Bond	*Bond energy*	*Bond*	*Bond energy*
C—H	415	C—Cl	328
C—C	344	C—Br	276
C—O	350	C—I	240
C—N	292	F—F	158
C—S	259	N—H	391
C—F	441	O—H	463
S—H	368	S—S	266
N—O	175	O—O	143
N—N	159	N—Cl	200
H—H	436	C═C	615
Cl—Cl	243	C═O	725
C═N	615	N═N	418
O═O	498	C≡C	812
C°N	890	N≡N	946

Table 10 Dipole moments (μ) of different bonds.

Bond	*μ*	*Bond*	*μ*
H—O	1.5 D	C—O	0.8 D
H—N	1.3 D	C═O	2.5 D
H—C	0.4 D	C—N	0.5 D
C—Cl	1.5 D	C≡N	3.5 D
C—Br	1.4 D		

Note: 1 D = 10^{-18} e.s.u.-cm. Dipole moment of F—C bond is assumed to be 1.0 D.

Table 11 Saturation vapour pressure table.

Temperature		*Saturation Vapour Pressure*	*Temperature*		*Vapour Pressure*
(°F)	*(°C)*	*(mbar)*	*(°F)*	*(°C)*	*(mbar)*
– 30	– 34.4	0.3	40	4.4	8.4
– 29	– 33.9	0.4	41	5.0	8.7
– 28	– 33.3	0.4	42	5.6	9.1
– 27	– 32.8	0.4	43	6.1	9.4
– 26	– 32.2	0.4	44	6.7	9.8
– 25	– 31.7	0.4	45	7.2	10.2
– 24	– 31.1	0.5	46	7.8	10.6
– 23	– 30.6	0.5	47	8.3	11.0
– 22	– 30.0	0.5	48	8.9	11.4
– 21	– 29.4	0.5	49	9.4	11.8
– 20	– 28.9	0.6	50	10.0	12.3
– 19	– 28.3	0.6	51	10.6	12.7
– 18	– 27.8	0.6	52	11.1	13.2
– 17	– 27.2	0.7	53	11.7	13.7
– 16	– 26.7	0.7	54	12.2	14.2
– 15	– 26.1	0.7	55	12.8	14.8
– 14	– 25.6	0.8	56	13.3	15.3
– 13	– 25.0	0.8	57	13.9	15.9
– 12	– 24.4	0.9	58	14.4	16.4
– 11	– 23.9	0.9	59	15.0	17.0
– 10	– 23.3	0.9	60	15.6	17.7
– 9	– 22.8	1.0	61	16.1	18.3
– 8	– 22.2	1.0	62	16.7	19.0
– 7	– 21.7	1.1	63	17.2	19.6
– 6	– 21.1	1.1	64	17.8	20.3
– 5	– 20.6	1.2	65	18.3	21.1
– 4	– 20.0	1.3	66	18.9	21.8
– 3	– 19.4	1.3	67	19.4	22.6
– 2	– 18.9	1.4	68	20.0	23.4
– 1	– 18.3	1.4	69	20.6	24.2
0	– 17.8	1.5	70	21.1	25.0
1	– 17.2	1.6	71	21.7	25.9
2	– 16.7	1.7	72	22.2	26.8

3	– 16.1	1.7	73	22.8	27.7
4	– 15.6	1.8	74	23.3	28.7
5	– 15.0	1.9	75	23.9	29.6
6	– 14.4	2.0	76	24.4	30.6
7	– 13.9	2.1	77	25.0	31.7
8	– 13.3	2.2	78	25.6	32.7
9	– 12.8	2.3	79	26.1	33.8
10	– 12.2	2.4	80	26.7	35.0
11	– 11.7	2.5	81	27.2	36.1
12	– 11.1	2.6	82	27.8	37.3
13	– 10.6	2.7	83	28.3	38.6
14	– 10.0	2.9	84	28.9	39.8
15	– 9.4	3.0	85	29.4	41.1
16	– 8.9	3.1	86	30.0	42.5
17	– 8.3	3.3	87	30.6	43.8
18	– 7.8	3.4	88	31.1	45.2
19	– 7.2	3.6	89	31.7	46.7
20	– 6.7	3.7	90	32.2	48.2
21	– 6.1	3.9	91	32.8	49.7
22	– 5.6	4.0	92	33.3	51.3
23	– 5.0	4.2	93	33.9	52.9
24	– 4.4	4.4	94	34.4	54.6
25	– 3.9	4.6	95	35.0	56.3
26	– 3.3	4.8	96	35.6	58.1
27	– 2.8	5.0	97	36.1	59.9
28	– 2.2	5.2	98	36.7	61.7
29	– 1.7	5.4	99	37.2	63.6
30	– 1.1	5.6	100	37.8	65.6
31	– 0.6	5.9	101	38.3	67.6
32	0.0	6.1	102	38.9	69.7
33	0.6	6.4	103	39.4	71.8
34	1.1	6.6	104	40.0	73.9
35	1.7	6.9	105	40.6	76.2
36	2.2	7.2	106	41.1	78.5
37	2.8	7.5	107	41.7	80.8
38	3.3	7.8	108	42.2	83.2
39	3.9	8.1	109	42.8	85.7
40	4.4	8.4	110	43.3	88.2

Appendix-II: Model Questions

Short Answer Type Questions (5 Marks each)

1. State the disadvantages of using *BOD* value.
2. Define environment. State different causes of environmental pollution.
3. Explain different types of abiotic components of environment.
4. State some objectives to study environmental science.
5. Define *COD*. State the advantages of *COD* over *BOD* for a given waste water sample.
6. State the biodegradation rate law. Explain with graph how *BOD* of a waste sample changes with time.
7. Explain eutrophication of a water reservoir. State some causes of eutrophication.
8. Discuss the necessary steps, which should be taken to prevent water pollution.
9. Define TLV of a pollutant. State its significance with examples.
10. BOD_5 of a waste sample is 200 *mg/l* and the ultimate *BOD* is 300 *mg/l*. Find the reaction rate constant. State the unit of the rate constant.
11. State some causes of water pollution.
12. Define carcinogen with three examples.
13. Discuss the harmful effects of hydrocarbons on the environment.
14. What are the six criteria pollutants as defined by NAAQS? Write down the permissible limits of them in the environment.
15. Define primary and secondary pollutants with examples.
16. (a) State the beneficial effects of ozone layer in the stratosphere.
    (b) "Tropospheric ozone is called bad ozone"—justify.
17. What is food chain? State three important characteristics of it. Hence define food web.
18. Explain the soil structure with a neat sketch.

19. Discuss the harmful effects of the following heavy metals: (i) As, (ii) Pb.
20. Discuss CO_2 cycle in the environment with a neat sketch.
21. The solar flux at the outer edge of the atmosphere of the Earth varies by ±3.3%. Calculate the change in equivalent temperature. Given: $\alpha = 0.3$.
22. Define acid rain and discuss its adverse effects on the environment.
23. Write a short note on nuclear fuel.
24. Define Threshold Limit Value of noise pollution. Explain how combined effect of noise pollution can be evaluated.
25. What is zero population growth? State the limitations of exponential growth model.
26. Discuss the benefits of implementing EIA.
27. Define the following terms: (i) demography, (ii) maximum sustainable yield.
28. Establish the following relationship using differential equation of logistic growth curve

$$r = \frac{R_0}{1 - \frac{N_0}{R}}$$

Where, R_0 = initial growth rate, r = average growth rate, N_0 = initial population size and K = carrying capacity.

29. What is oxygen depletion? State two harmful effects of sewage.
30. Define the following: (i) ventilation coefficient and (ii) mixing depth.
31. State the harmful effects of UV radiations.
32. State the difference between *DO* and *BOD*.
33. Discuss the steps to be taken to conserve biodiversity.
34. Explain the term "conservation status rank".
35. What is "Club of Rome"? What was the principal agenda of Club of Rome?
36. State some environmental problems in India.

Long Answer Type Questions (15 Marks each)

1. (a) Define the terms: (i) aerobic decomposition and (ii) anaerobic decomposition.
   (b) State the significances of dissolved oxygen (*DO*).
   (c) State the factors influencing *DO* level of a water reservoir.
   (d) Define pollutant. State three different water pollutants.

2. (a) Define *BOD* of a given waste water sample. State its unit.
   (b) Sate the significance of *BOD* and hence explain how quality of a waste water sample can be quantified.
   (c) Discuss the reagents required in determination of *COD* and the function of each reagent.
   (d) State causes of eutrophication. How can it be controlled?
3. (a) Discuss different stages of hydrological cycle.
   (b) Define aquifer and classify it.
   (c) Classify biotic and abiotic components of an ecosystem. Give examples of each category.
4. (a) Explain primary treatment process of waste water treatment.
   (b) State some disadvantages of trickling filter method.
   (c) What is called *activated sludge*? Why is it called so?
   (d) Explain the function of dry lime and bleaching powder in waste water treatment process.
5. (a) Explain (i) oxygen deficit and (ii) reaeration rate constant.
   (b) State some characteristic features of ecosystem.
   (c) What are heavy metals? Give three examples.
6. (a) What is photochemical smog? What are the constituents of photochemical smog?
   (b) State the differences between photochemical smog and reducing smog.
   (c) Write the full form of SPM. What is aerosol?
   (d) State the harmful effects of aerosol on the environment.
7. (a) Discuss the harmful effects of the following heavy metals: (i) Cd, (ii) Hg.
   (b) Discuss the harmful effects of the following on biotic component: (i) CO, (ii) SO_2, (iii) NO_2.
8. (a) What is PAH? Write three examples. What is soot particle?
   (b) Write down the sources of PAH. State the harmful effects of PAH.
9. (a) What is an ESP? Discuss its utility in controlling air pollution.
   (b) Explain the principles of filter bag unit and cold chamber in exhaust air treatment.
   (c) An electrostatic precipitator with 6000 m^2 of collector plate area is 97% efficient in treating 200 m^3/s of flue gas from a 200 MW power plant. Calculate the required plate area to increase the efficiency from 97% to 98% and to 99%?

10. (a) What is biogeochemical cycle?
    (b) Briefly discuss the following cycles with a neat sketch: (i) Nitrogen cycle, (ii) Sulphur cycle, (iii) Phosphorus cycle.

11. (a) Define noise. Explain A-scale and B-scale of noise level.
    (b) Calculate the combined loudness of two noise levels of 60 dBA and 80 dBA.

12. (a) Explain ODS with examples.
    (b) Explain how CFC decreases the ozone concentration in the stratosphere.
    (c) Define ODP. State the unit of ODP.
    (d) Write a short note on Montreal Protocol.

13. (a) Illustrate environmental strain and environmental stress with examples.
    (b) Explain the terms: GPP and NPP.
    (c) State the factors influencing the GPP.
    (d) Draw a neat sketch of energy flow diagram in different biotic components.

14. (a) Discuss the adverse effects of global warming. Define GWP.
    (b) Discuss the remedial measures to be taken to control global warming.
    (c) Consider the Earth is flat and an isothermal body. Albedo for the Earth is 0.2. Estimate the temperature of the flat Earth.
    Given: $S = 1372$ W/m², $\sigma = .67 \times 10^{-8}$ W/m²(K)⁴

15. (a) What is albedo? What is its value for the Earth?
    (b) State Wien's displacement law.
    (c) Define Greenhouse Effect. Name four major GHGs.
    (d) State the beneficial effects of Greenhouse Effect.

16. (a) Draw a neat sketch of atmosphere, showing different layers.
    (b) Define adiabatic lapse rate and ambient lapse rate and hence explain atmospheric stability.

17. (a) What is solid waste? Classify it with examples.
    (b) Briefly explain different treatment processes of solid wastes.
    (c) State the causes of soil pollution.
    (d) Discuss the steps to be taken to prevent soil pollution.

18. (a) Explain why composting is the best suitable method for disposal of biodegradable MSW.
    (b) State the adverse effects of open dumping of MSW on environment.

19. (a) Briefly discuss the sanitary landfill method of MSW. How it differs from open dumping?
    (b) State different components of a cell in sanitary landfill method.

20. (a) Explain the following with examples: (i) Contaminants, (ii) Receptor, (iii) Speciation.
    (b) Explain how pollution due to automobile exhaust can be controlled.
    (c) Explain the term PSI.

21. (a) Solar flux or insolation (S), striking on a planet, is given by:

$$s \propto \frac{1}{d^2} = \frac{k}{d^2}$$

, where, ds is the distance between the Sun and the planet.

   (i) Find k, given d for the Earth is 150×10^9 m and S = 1372 W/m².
   (ii) Estimate d for the Venus. Given: $S = 2613$ W/m²
   (iii) Estimate S for the Mercury. Given: $d = 58 \times 10^9$ m

   (b) Estimate equivalent temperature and surface temperature of the Mars by using the following data:
   $S = 589$ W/m₂, $\alpha = 0.15$, $\sigma = 5.67 \times 10^{-8}$ W/m² (K)⁴, $\lambda_{max} = 13.173$ μm
   (c) Suppose world energy consumption of fossil fuel is 3×10^{17} kJ/year, which is assumed to be obtained by combustion of petroleum with approximate formula C_2H_3. Combustion of petroleum releases about 43×10^3 kJ/kg. Estimate the emission rates of CO_2 and C in the atmosphere.

22. (a) Define doubling time (T_d) in an exponential population growth. Show that

$$T_d = \frac{69.3}{r}$$

   Where, r is the percent population growth rate constant and ln 2 = 0.693.
   (b) Write down the expression of logistic model of population growth. In 1970, the world's population was 4 billion and growth rate was 2% per year. Steady-state population is 12 billion. When would the population reach 6 billion? What would be the projected population in 1995 and 2025 using logistic model?
   (c) Define: (i) carrying capacity and environmental resistance.

23. (a) Deduce the following relation $BOD_t = L_0(1 - e^{-k_d t})$, where symbols have usual significances.
   (b) What is called *NBOD* and *TBOD*? Discuss how *NBOD* of a given waste water sample can be calculated?
   (c) Find out the ratio of BOD_5 at 20°C to that of $BOD_{2.5}$ at 35°C.

24. A *BOD* test is run using 50 ml of waste water mixed with 100 ml of pure water. The initial *DO* of the mixture is 8 *mg/l*. After a long time the *DO* remains fixed at 1 *mg/l*.
    (a) What is the *BOD* of the waste water?
    (b) What is the ultimate *BOD*?
    (c) What is the remaining *BOD* after 5 days?
    (d) What is the reaction rate constant at 20ºC?
    (e) Find out the value of rate constant at 45ºC.

25. (a) What is called atmospheric stability?
    (b) Explain (i) superadiabatic ALR and (ii) subadiabatic ALR.
    (c) When the atmosphere is called neutrally stable and why?

26. (a) Write down the different components in an aquatic ecosystem.
    (b) Discuss the Winkler method of analysis of *DO* in the laboratory.
    (c) What is azide modification of basic Winkler method?

27. (a) Define and classify biodiversity.
    (b) Define biodiversity hotspots with example.
    (c) State the causes of loss of biodiversity.

28. (a) Define diversity index and hence explain richness factor and evenness factor with examples.
    (b) Explain the terms (i) Simpon's index and (ii) Shannon index.

29. (a) Define sustainable development and explain why it is necessary.
    (b) State the basic concepts of sustainable development.

Multiple Choice Questions (1 Mark each)

Choose the correct answer/alternative of the following.

1. PAN is a
   (a) constituent of photochemical smog,
   (b) carcinogen,
   (c) heavy metal,
   (d) GHG.

2. Which one of the following is not a GHG?
   (a) CO_2, (b) N_2,
   (c) N_2O, (d) CH_4.

3. The value of albedo of the Earth is
   (a) 0.21, (b) 0.31,
   (c) 0.41, (d) 0.51.

4. Addition of two sound levels, 100 dB each, gives the resultant sound level of
   (a) 100 dB, (b) 200 dB,
   (c) 103 dB, (d) 53 dB.

5. Which one of the following is an example of occupational noise?
   (a) Road traffic noise, (b) Lathe machine noise,
   (c) Aircraft noise, (d) Loudspeaker noise.

6. Maximum temperature of stratosphere is
   (a) –95°C, (b) 0°C,
   (c) 52°C, (d) –56°C.

7. Which one of the following is an example of secondary pollutant?
   (a) SO_2, (b) NO_2,
   (c) CO, (d) H_2SO_4.

8. The full form of EIA is
   (a) Electric Intensity Assessment, (b) Environmental Impact Act,
   (c) Environmental Impact Agencies, (d) Environmental Impact Assessment.

9. Which one of the following attacks hemoglobin?
   (a) CO, (b) NO_2,
   (c) O_3, (d) CO_2.

10. The full form of PAH is
    (a) Polycyclic Aliphatic Hydrocarbon,
    (b) Phenolic Aromatic Hydrocarbon,
    (c) Polyhydric Aromatic Hydrocarbon,
    (d) Phenolic Aliphatic Hydrocarbon.

11. Which one of the following is the worst enemy of ozone?
    (a) CFC, (b) NO,
    (c) Pb, (d) SO_2.

12. The hottest region of the atmosphere is
    (a) Troposphere, (b) Mesosphere,
    (c) Stratosphere, (d) Ionosphere.

13. The relation between GPP and NPP is
    (a) GPP = NPP, (b) GPP < NPP,
    (c) GPP + R = NPP, (d) GPP = NPP + R.

14. The full form of PSI is
   (a) Polymer Standard Index, (b) Polystyrene index,
   (c) Pollution Standard Index, (d) Primary Standard Index.

15. Which one of the following is a renewable resource?
   (a) Coal, (b) Iron,
   (c) Oil, (d) Water.

16. Which one of the following is a nephrotoxic element?
   (a) Fe, (b) Ag,
   (c) Hg, (d) Cu.

17. The expression of dry adiabatic lapse rate is
   (a) $-\frac{C_P}{g}$, (b) $-\frac{\rho C_P}{g}$,
   (c) $-\frac{g}{\rho C_P}$, (d) $-\frac{g}{C_P}$.

18. Kyoto protocol is related with
   (a) Ozone layer depletion, (b) Greenhouse Gas,
   (c) Air pollution, (d) Water pollution.

19. The best method of solid waste disposal is
   (a) incineration, (b) composting,
   (c) landfill, (d) shredding.

20. Montreal Protocol is related with
   (a) Water pollution, (b) SPM,
   (c) Greenhouse Gas, (d) Ozone.

21. Catalytic converter is used to reduce
   (a) automobile exhaust pollution, (b) factory exhaust pollution,
   (c) CO_2 emission, (d) N_2 emission.

22. In air pollution control cyclone separator is used to
   (a) absorb harmful hydrocarbons,
   (b) separate volatile organic compounds,
   (c) remove colloidal size particles,
   (d) remove large size SPM.

23. The road traffic noise is measured by
   (a) L_{10} (18 hours) index, (b) Lepn,
   (c) L_{eq}, (d) dBA.

24. In stratosphere the value of dry adiabatic lapse rate is
   (a) positive, (b) negative,
   (c) 0, (d) 1.

25. Which one of the following is called haematoxin?
   (a) Cu, (b) Pb,
   (c) As, (d) Hg.

26. Methyl isocyanate is a
   (a) primary pollutant, (b) contaminant,
   (c) secondary pollutant, (d) criteria pollutant.

27. The unit of intensity level of noise is
   (a) meter, (b) Newton,
   (c) candle, (d) deciBel.

28. Loudness of noise level is given by:
   (a) $\ln\frac{I}{I_0}$, (b) $10\ln\frac{I}{I_0}$,
   (c) $10\log\frac{I_0}{I}$, (d) $10\log\frac{I}{I_0}$.

29. The average surface temperature of Earth is
   (a) 25°C, (b) 10°C,
   (c) 15°C, (d) 20°C.

30. Which one of the following is the main component of mesosphere?
   (a) NO, (b) H_2^+,
   (c) N_2, (d) CO_2.

31. Which one of the following heavy metals was responsible for Minamata episode?
   (a) Pb, (b) As,
   (c) Cd, (d) Hg.

32. Which one of the following is considered as a source of underground water?
   (a) Ocean, (b) Aquifer,
   (c) Marsh, (d) Pond.

33. The threshold limit value for 8 hours is
   (a) 80 dB, (b) 85 dB,
   (c) 90 dB, (d) 95 dB.

34. Lapse rate refers to the rate at which
   (a) change in pressure with increase in altitude,
   (b) change in volume with increase in temperature,

(c) change in pressure with increase in temperature,
(d) change in temperature with increase in altitude.

35. Which one of the following is not a constituent of acid rain?
(a) Acetic acid, (b) Sulfuric acid,
(c) Hydrochloric acid, (d) Nitric acid.

36. To avoid eutrophication of an oligotrophic lake N/P ratio should be maintained
(a) over 20, (b) below 5,
(c) around 1, (d) around 7.

37. The term GWP is related with
(a) Ozone layer depletion, (b) Hydrocarbon emission,
(c) Greenhouse Gas, (d) Pb particle emission.

38. Which one of the following is the chemical formula of CFC – 115?
(a) $C_3H_2F_4Cl_2$, (b) C_2HF_4Cl,
(c) $C_3F_4Cl_4$, (d) C_2F_5Cl.

39. Which one of the following can be used for disinfection of water?
(a) UV-ray, (b) Hydrogen peroxide,
(c) Ozone, (d) Carbon dioxide.

40. The pollutant ozone is present in
(a) Stratosphere, (b) Mesosphere,
(c) Troposphere, (d) Thermosphere.

41. The most commonly used catalyst in catalytic converters is finely divided
(a) Iron, (b) Palladium,
(c) Platinum, (d) Nickel.

42. Energy flow of an ecosystem is
(a) non-directional, (b) bi-directional,
(c) cyclic, (d) unidirectional.

43. Which one of the following is responsible for permanent hardness of water?
(a) Chlorides, (b) Carbonates,
(c) Bicarbonates, (d) Phosphates.

44. Biomagnification is related to the accumulation of
(a) biodegradable substances, (b) non-biodegradable substances,
(c) chemical substances, (d) minerals.

45. Which of the following energy is eco-friendly?
(a) Fossil fuel, (b) Hydropower,
(c) Nuclear power, (d) Wood.

46. Petroleum is a
    (a) renewable resource,
    (b) non-renewable resource,
    (c) Fixed resource,
    (d) non-polluting resource.

47. Keeping and nurturing of forest is called
    (a) siviculture,
    (b) agriculture,
    (c) horticulture,
    (d) floriculture.

48. Which of the following is not a biotic component?
    (a) Herbivores,
    (b) Decomposers,
    (c) Carnivores,
    (d) Minerals.

49. Tropical rainforest is found in
    (a) Gangetic plain land,
    (b) Deccan plateau,
    (c) Western Himalayas,
    (d) Western ghats.

50. Which of the following is a part of detritus in an ecosystem?
    (a) Dead plant parts,
    (b) Dead animals,
    (c) Animal excreta,
    (d) All of these.

51. Bacteria and fungi are called
    (a) herbivores,
    (b) carnivores,
    (c) consumers,
    (d) decomposers.

52. Ozone depletion will cause
    (a) increased incidence of skin cancer,
    (b) forest fires,
    (c) global warming,
    (d) increase in SPM.

53. For any pollutant
    (a) higher the TLV, the pollutant is said to have more toxic effect,
    (b) higher the TLV, the pollutant is said to have less toxic effect,
    (c) lower the TLV, the pollutant is said to have less toxic effect,
    (d) TLV is no way connected with toxic effect.

54. The critical *DO* for any water reservoir is
    (a) 6 *mg/l*,
    (b) 4 *mg/l*,
    (c) 10 *mg/l*,
    (d) 15 *mg/l*.

55. Which one of the following is true for a given water sample?
    (a) *COD* is always less than *BOD*,
    (b) *COD* cannot be related with *BOD*,
    (c) *BOD* value is equal to *COD* value,
    (d) *BOD* is always less than *COD*.

56. *BOD* test is performed at dark
    (a) to prevent gases to diffuse,
    (b) to prevent bacteria to enter into the bottle,
    (c) to accelerate biodegradation,
    (d) to prevent light to enter into the bottle.

57. In presence of nephrotoxin, which one of the following is affected?
    (a) blood, (b) liver,
    (c) kidney, (d) lungs.

58. With increase in temperature the *DO* content of water
    (a) increases, (b) decreases,
    (c) remains constant, (d) initially increases then decreases.

59. Which one of the following should contain maximum *DO*?
    (a) molten Glacier and ice caps, (b) River,
    (c) Lake, (d) Ground water.

60. Which one of the following is the cause of eutrophication?
    (a) Increase of pathogens, (b) Increase of *BOD*,
    (c) Increase of algae's productivity, (d) Increase of *DO*.

61. Temporary hardness of water is due to the presence of
    (a) Cl^-, (b) SO_4^{2-},
    (c) HCO_3^-, (d) NO_3^-.

62. Blue baby syndrome is related to
    (a) NO_2^-, (b) PO_4^{3-},
    (c) NO_3^-, (d) SO_4^{2-}.

63. "Minamata disease" is associated with
    (a) arsenic, (b) mercury,
    (c) cadmium, (d) lead.

64. Ozonized water means
    (a) water containing ozone, (b) ozone added as disinfectant,
    (c) deionized water, (d) saline water.

65. The most important elements causing algal bloom are
    (a) C, N, P, (b) Ca, Mg, Co,
    (c) Mo, Fe, Cu, (d) N, P, K.

66. Atmosphere is said to be unstable under the condition of
    (a) ELR > ALR, (b) ELR < ALR,
    (c) ELR = ALR, (d) ELR ~ ALR.

67. London smog is also known as
(a) nitrogenous smog, (b) sulphurous smog,
(c) phosphorus smog, (d) chlorinated smog.

68. Ozone layer thickness is expressed in
(a) decibel, (b) cm,
(c) nm, (d) Dobson unit.

69. Which one of the following is a primary pollutant?
(a) PAH, (b) SO_2,
(c) Cl_2, (d) PAN.

70. The hottest region of the atmosphere is
(a) stratosphere, (b) troposphere,
(c) thermosphere, (d) mesosphere.

71. The main constituent of photochemical smog is
(a) H_2O, (b) SO_2,
(c) NO_2, (d) O_3.

72. For air stability, which one of the following is correct?
(a) DALR = ELR, (b) DALR > ELR,
(c) DALR < ELR, (d) DALR >> ELR.

73. Sound that is safe to the human ear should not exceed
(a) 45 dB, (b) 125 dB,
(c) 70 dB, (d) 85 dB.

74. Which one of the following attacks blood cell?
(a) Cd, (b) As,
(c) Pb, (d) Hg.

75. MSW are disposed safely in
(a) caves, (b) sea,
(c) landfills, (d) open air.

76. The best source of phosphorus in an ecosystem is
(a) plants, (b) animals,
(c) soil, (d) rock.

77. A biome is characterized by
(a) vegetation, (b) biological communities,
(c) habitats, (d) environment.

78. Agenda 21 is related to
(a) Greenhouse gases, (b) sustainable development,
(c) biodiversity, (d) nuclear pollution.

Bibliography

1. A. Mallick, *Chemistry for Engineers*, Viva Books Pvt. Ltd. (2012).
2. Gilbert, M. Masters, *Introduction to Environmental Engineering and Science (2nd Edition)*, Pearson Education (2004).
3. R.A. Corbitt, *Handbook of Environmental Engineering*, McGraw-Hill, New York (1990).
4. A. Newman, *The Greening of Environmental Labelling, Environmental Science and Technology*, 25, 12 (1991).
5. E.A. McBean, F.A. Rovers and G.J. Farquhar, *Solid Waste Landfill Engineering and Design*, Prentice Hall, Englewood Cliffs, NJ, USA (1995).
6. C.R. Rhyner, L.J. Schwartz, R.B. Wenger and M.G. Kohrell, *Waste Management and Resource Recovery*, Lewis Publishers, New York (1995).
7. Anil K. De, *Environmental Chemistry (7th Edition)*, New Age International Publishers (2010).
8. T.G. Spiro and W. M. Stigliani, *Environmental Issues in Chemical Perspectives*, State University of New York, Albany, USA (1980).
9. S.E. Manahan, *Environmental Chemistry (6th Edition)*, CEC Press, USA (1994).
10. S.D. Lee, *Biochemical Effects of Environmental Pollutants*, Ann Arbor Science Publishers Inc., Ann. Arbor, Mich (1977).
11. S.E. Manahan, *Toxicological Chemistry (2nd Edition)*, Lewis Publishers, Chelsea (1992).
12. D.F.S. Natusch and P.K. Hopke, *Analytical Aspects of Environmental Chemistry*, John Wiley & Sons, New York, USA (1983).
13. B. Joseph, *Environmental Studies*, Tata McGraw Hill Publishing Co. Ltd., New Delhi, India (2005).

14. R.H. Whittaker, *Communities and Ecosystems (2nd Edition)*, Collier Macmillan, London, UK (1975).

15. E.J. Kormondy, *Concepts of Ecology (4th Edition)*, Prentice Hall of India Pvt. Ltd., New Delhi, India (2003).

16. V.H. Heywood, *Global Biodiversity Assessment*, Cambridge University Press, UNEP, Cambridge, UK (1995).

17. J.L. Chapman and M.J. Reiss, *Ecology: Principles and Applications*, Cambridge University Press, Cambridge, UK (1992).

18. D. Botbin and E. Keller, *Environmental Science*, John Wiley & Sons Inc. USA (1995).

19. W.P. Cumingham and B.W. Saigo, *Environmental Science*, McGraw Hill USA (1997).

20. S. Deswal and A. Deswal, *Energy, Ecology, Environment & Society*, Dhanpat Rai & Co. (P) Ltd., Delhi (2002).

21. H. Jadhav and V.M. Bhosale, *Environmental Protection and Laws*, Himalaya Pub. House, Delhi (1995).

22. E.P. Odum, *Fundamentals of Ecology*, W.B. Saunders Co. USA (1971).

23. H.S. Peavy et.al., *Environmental Engineering*, McGraw Hill Book Company, N.Y. (1986).

24. K.D. Wagner, *Environmental Management*, W.B. Saunders Co., Philadelphia, USA (1998).

Index

O

P